ADVENTURES IN THE MUSCULAR SYSTEM

THE BOD SQUAD

by Alexander Lowe • illustrated by Sebastian Kadlecik

NORWOOD DISCOVERY Graphics

Norwood House Press

For more information about Norwood House Press please visit our website at: www.norwoodhousepress.com or call 866-565-2900.

Copyright ©2021 by Norwood House Press. All rights reserved.

No part of this book may be reproduced or utilized in any form or by any means without written permission from the publisher.

Library of Congress Cataloging-in-Publication Data
Names: Lowe, Alexander, author. | Kadlecik, Sebastian, illustrator.
Title: Adventures in the muscular system / Alexander Lowe ; illustrated by Sebastian Kadlecik.
Description: Chicago : Norwood House Press, 2020. | Series: Norwood discovery graphics | Audience: Ages 8-10 |
  Audience: Grades 4-6 | Summary: "As Logan jumps on the bed, The Bod Squad sets out to explore his muscular system. The squad shrinks down in size, traveling and exploring how his muscles and bones react while also helping him avoid a fall. An adventure-filled graphic novel that provides information about the human body and how its muscular system works. Includes contemporary full-color graphic artwork, fun facts, additional information, and a glossary"— Provided by publisher.
Identifiers: LCCN 2020024502 (print) | LCCN 2020024503 (ebook) | ISBN 9781684508594 (hardcover)
  | ISBN 9781684045839 (paperback) | ISBN 9781684045884 (epub)
Subjects: LCSH: Musculoskeletal system—Comic books, strips, etc. | Musculoskeletal system—
  Juvenile literature. | Muscles—Juvenile literature. | Graphic novels.
Classification: LCC QM151 .L69 2020 (print) | LCC QM151 (ebook) | DDC 612.7—dc23
LC record available at https://lccn.loc.gov/2020024502
LC ebook record available at https://lccn.loc.gov/2020024503

Hardcover ISBN: 978-1-68450-859-4   Paperback ISBN: 978-1-68404-583-9

328N—072020
Manufactured in the United States of America in North Mankato Minnesota.

Contents

Story 4

The Muscular System 30

Glossary 31

Further Reading 31

About the Author 32

Meet the Bod Squad

Jada

Kara

Logan

Sam

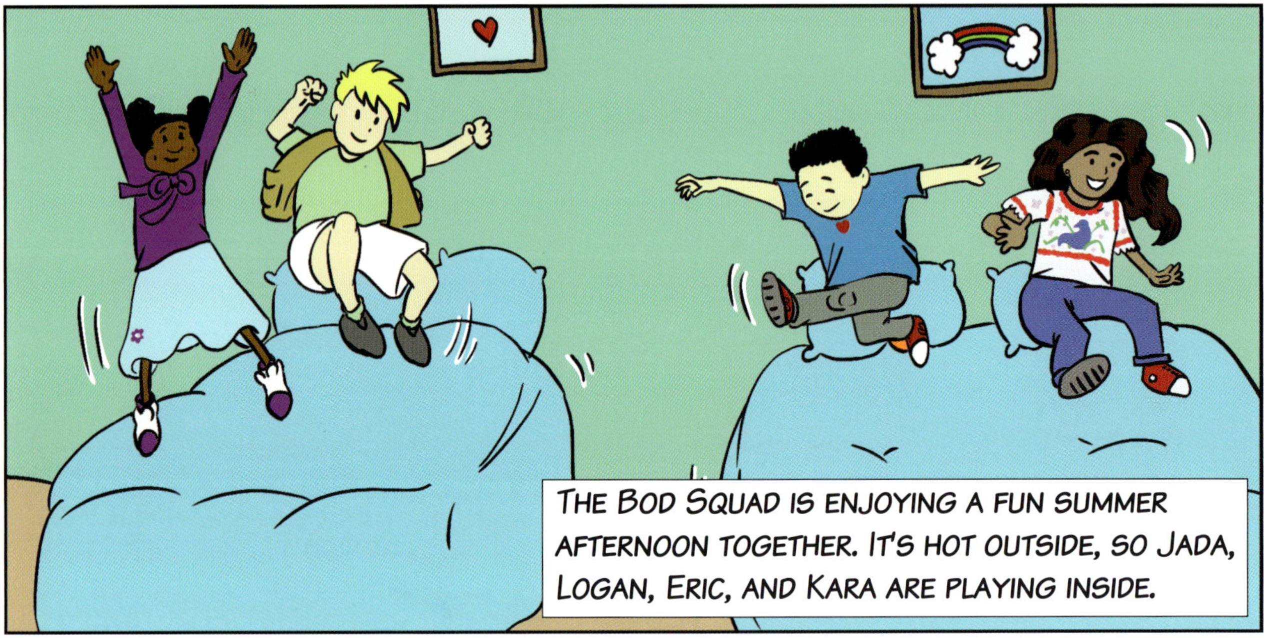
THE BOD SQUAD IS ENJOYING A FUN SUMMER AFTERNOON TOGETHER. IT'S HOT OUTSIDE, SO JADA, LOGAN, ERIC, AND KARA ARE PLAYING INSIDE.

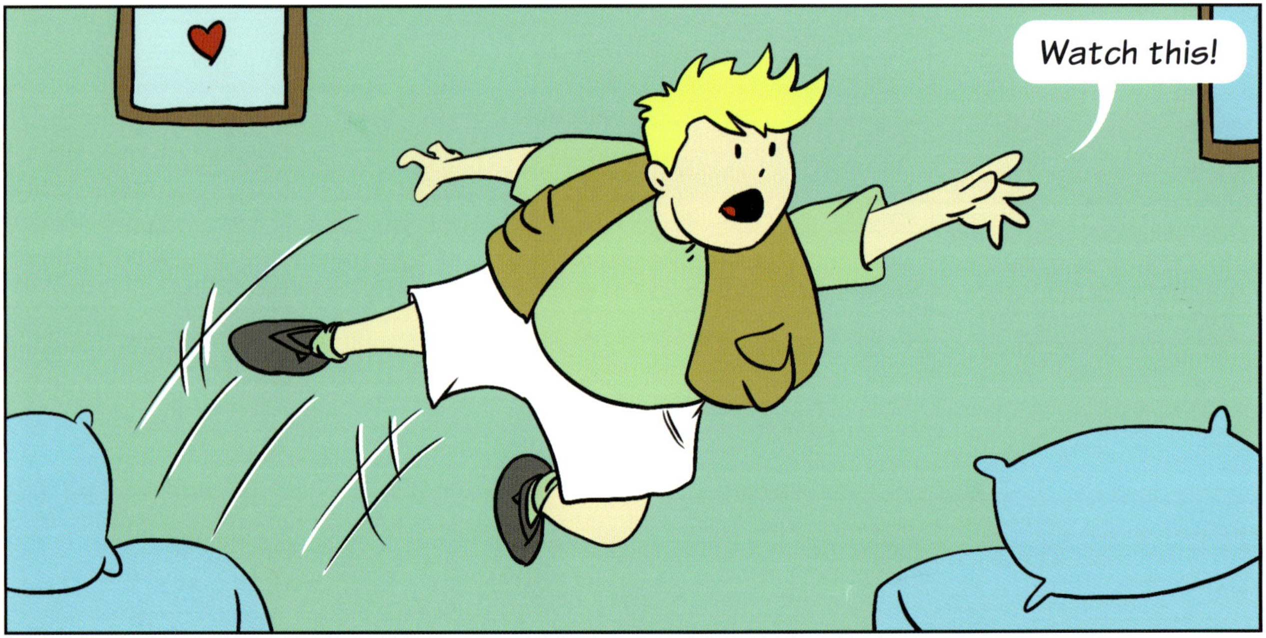
Watch this!

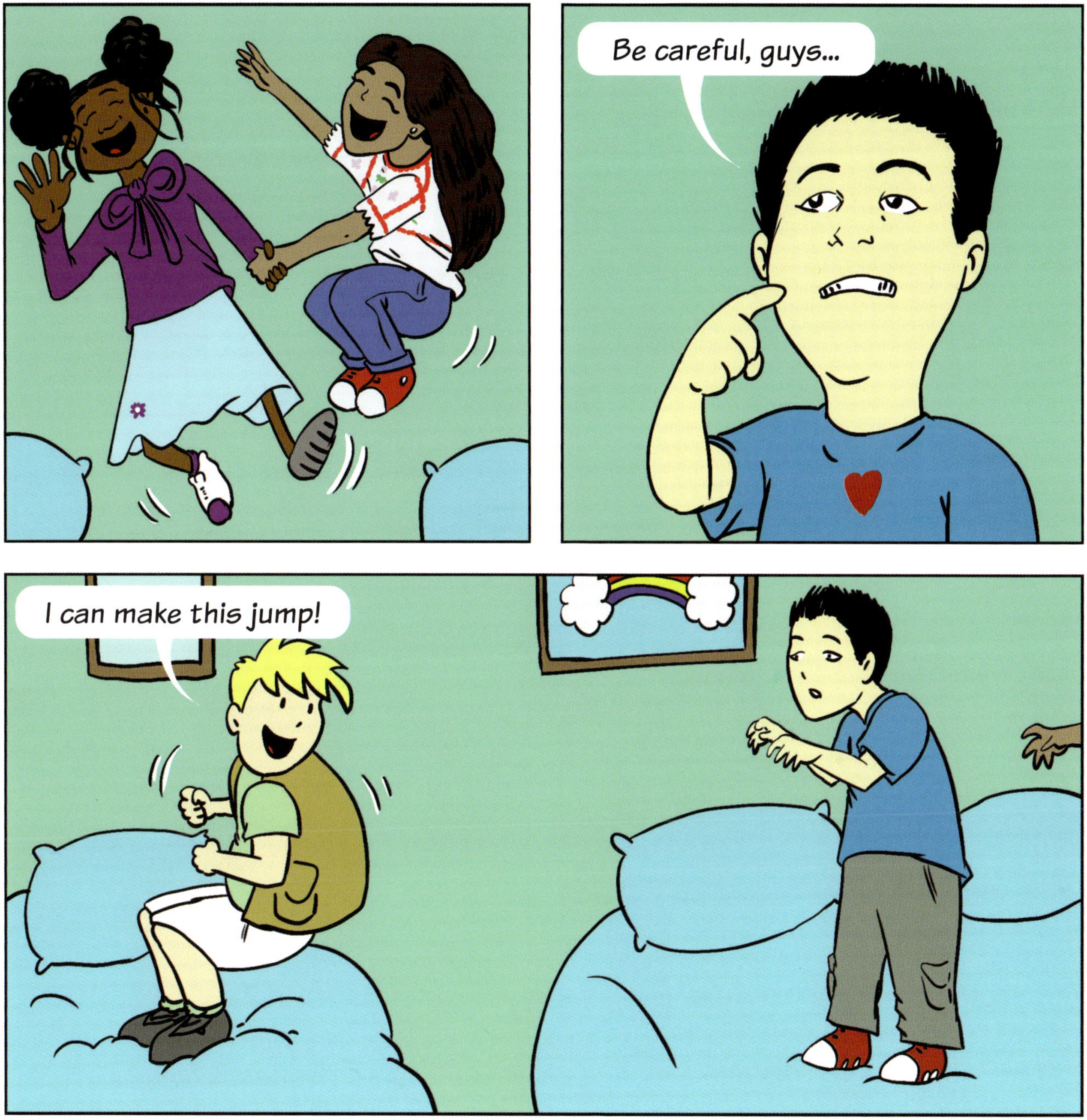
Be careful, guys...
I can make this jump!

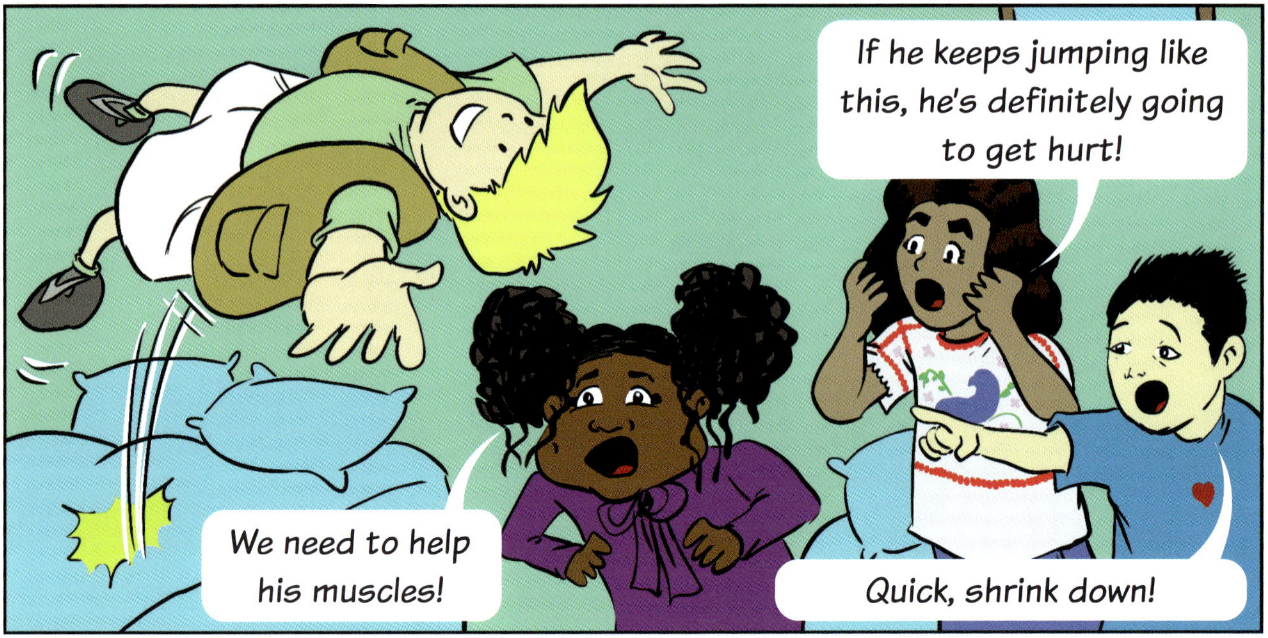
If he keeps jumping like this, he's definitely going to get hurt!
We need to help his muscles!
Quick, shrink down!

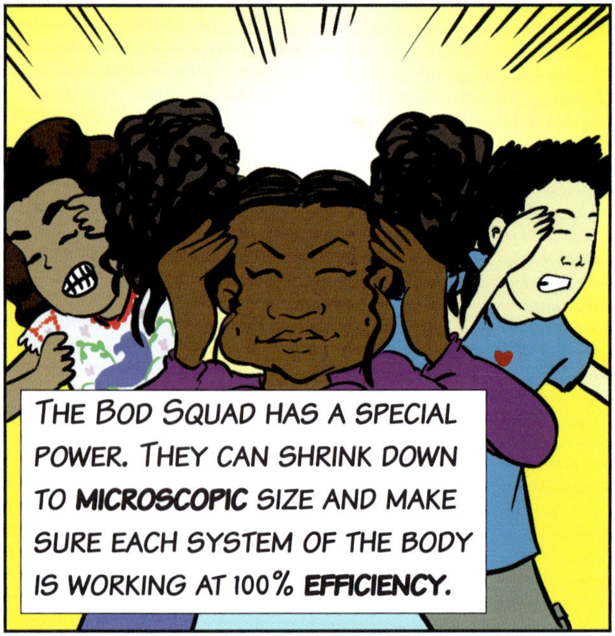
THE BOD SQUAD HAS A SPECIAL POWER. THEY CAN SHRINK DOWN TO **MICROSCOPIC** SIZE AND MAKE SURE EACH SYSTEM OF THE BODY IS WORKING AT 100% **EFFICIENCY**.

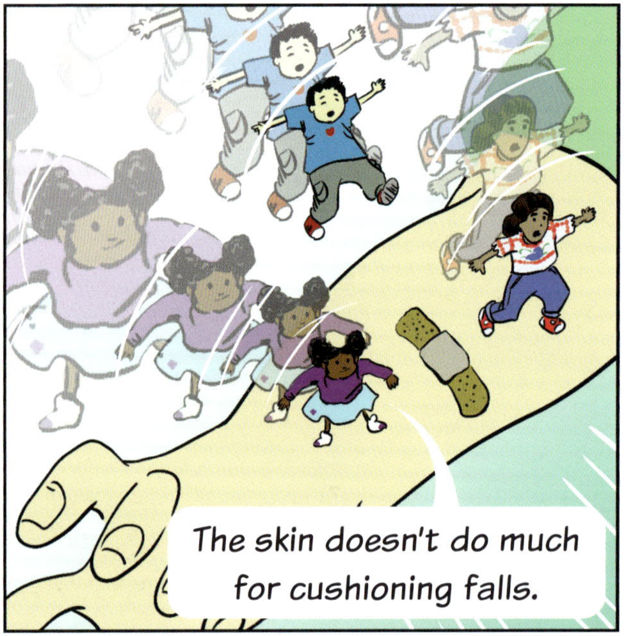
The skin doesn't do much for cushioning falls.

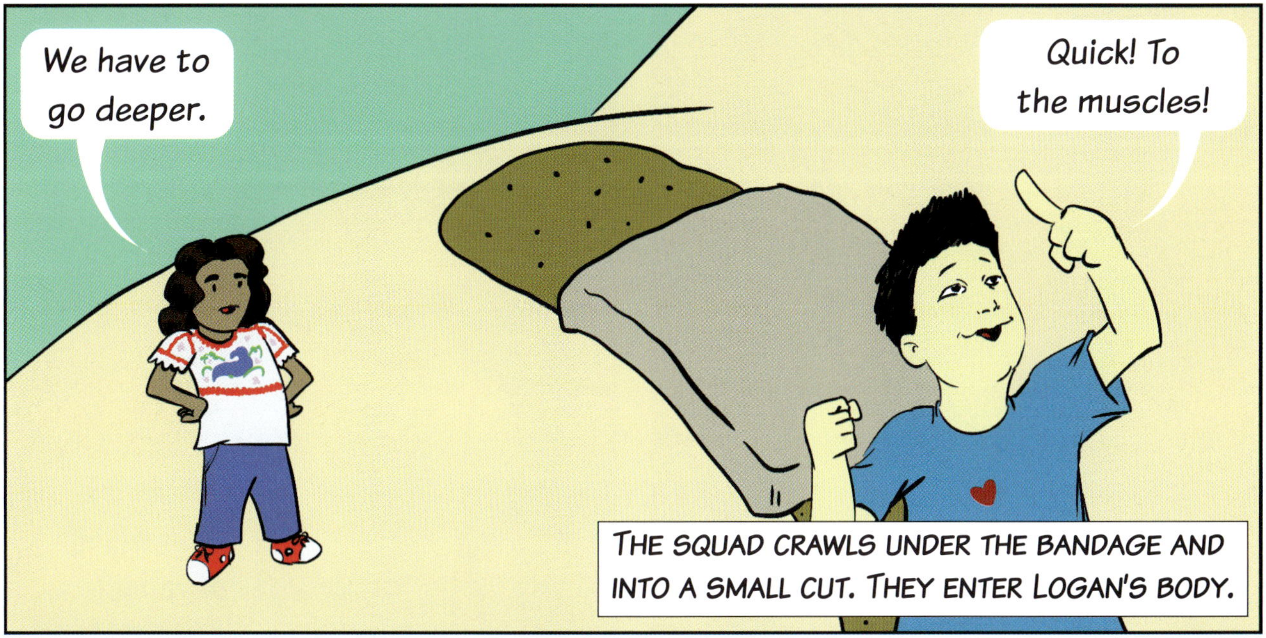
We have to go deeper.
Quick! To the muscles!
THE SQUAD CRAWLS UNDER THE BANDAGE AND INTO A SMALL CUT. THEY ENTER LOGAN'S BODY.

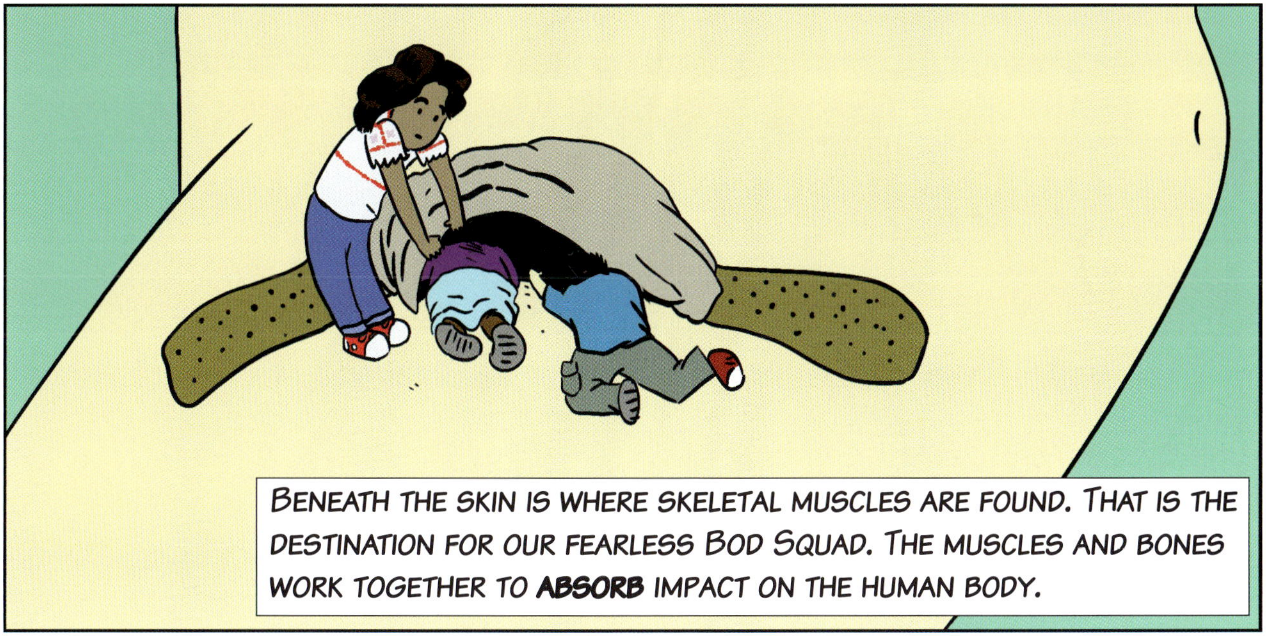
BENEATH THE SKIN IS WHERE SKELETAL MUSCLES ARE FOUND. THAT IS THE DESTINATION FOR OUR FEARLESS BOD SQUAD. THE MUSCLES AND BONES WORK TOGETHER TO **ABSORB** IMPACT ON THE HUMAN BODY.

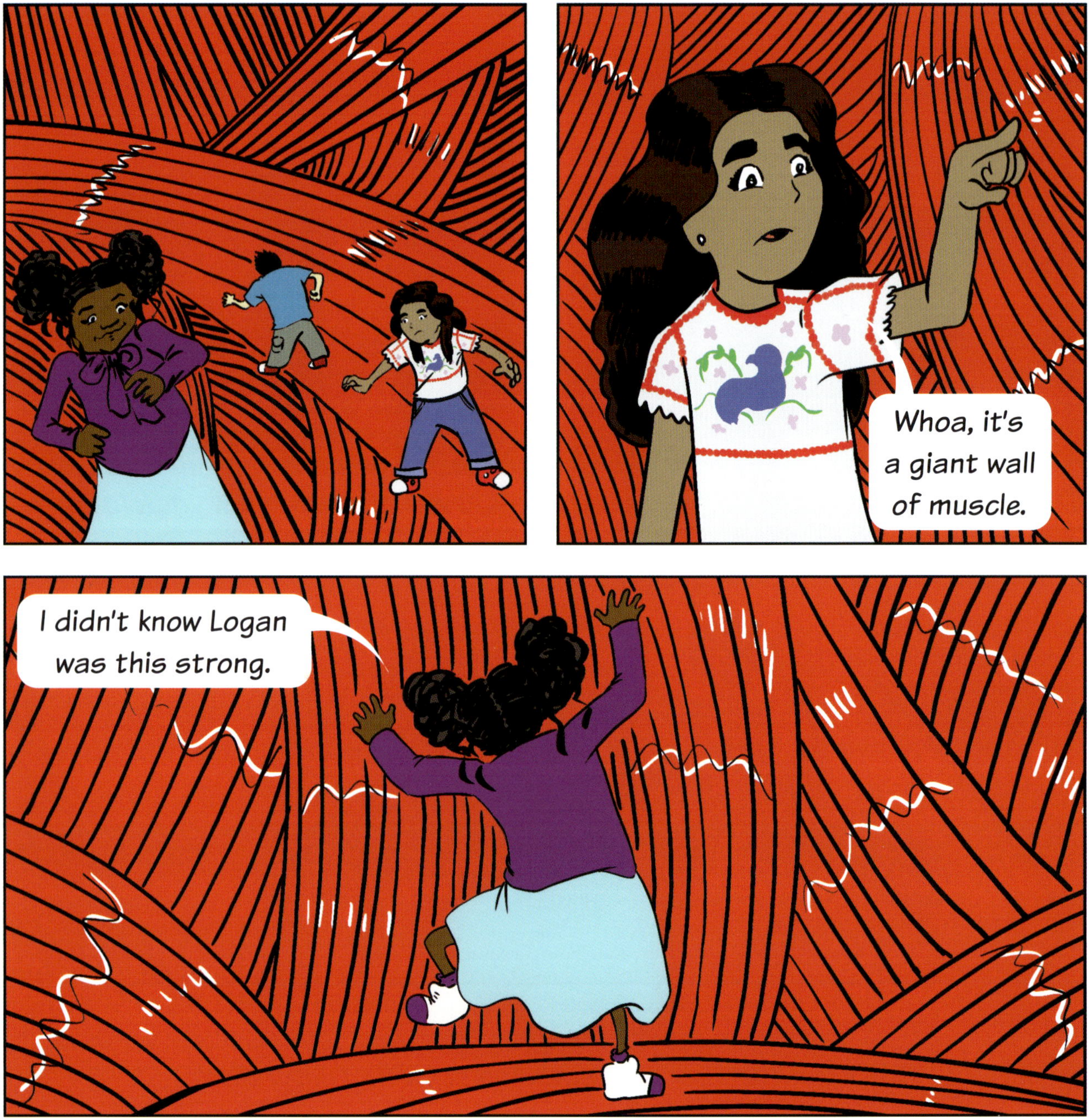
Whoa, it's a giant wall of muscle.
I didn't know Logan was this strong.

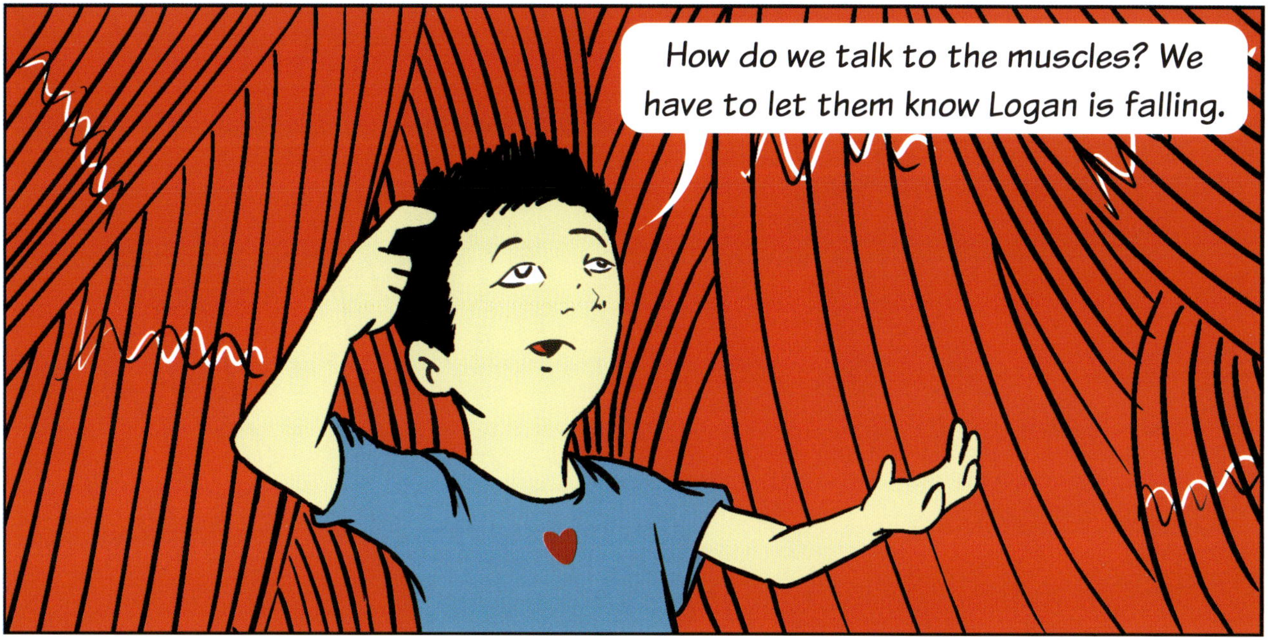
How do we talk to the muscles? We have to let them know Logan is falling.

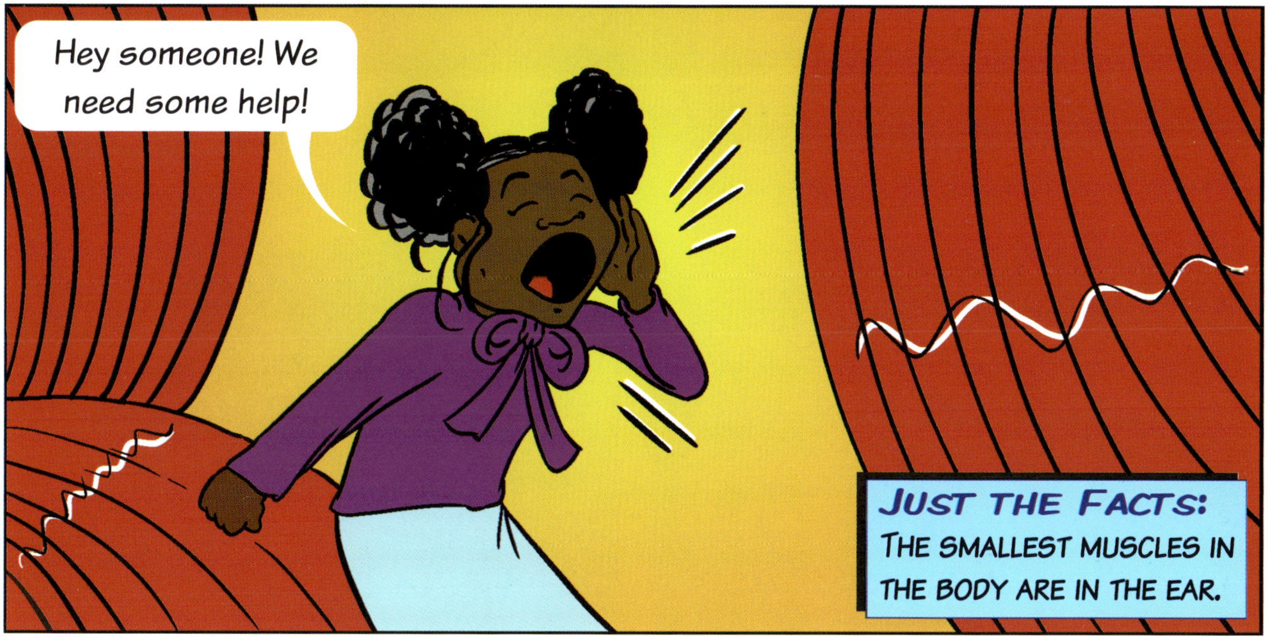
Hey someone! We need some help!
JUST THE FACTS:
THE SMALLEST MUSCLES IN THE BODY ARE IN THE EAR.

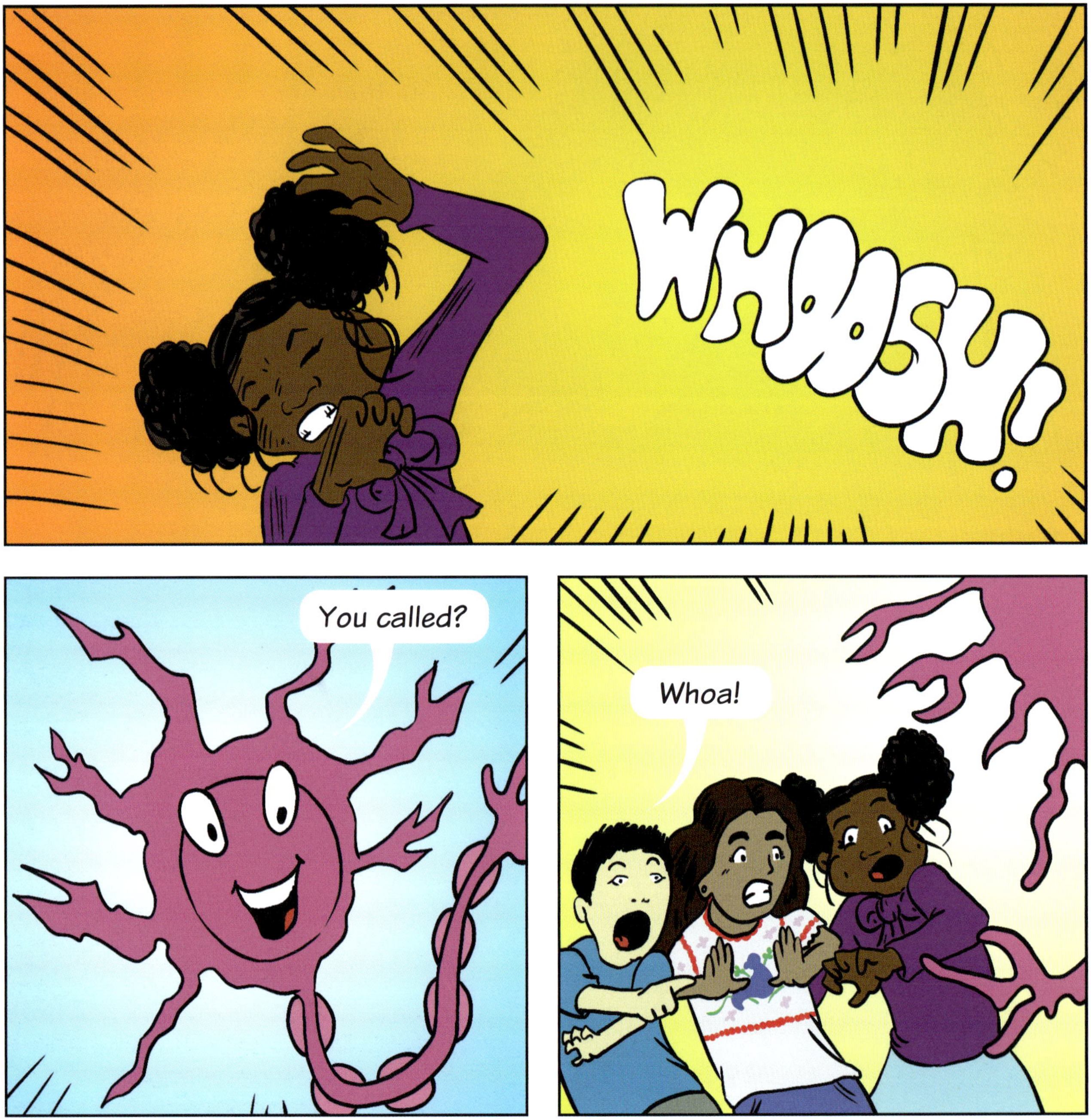
WHOOSH!
You called?
Whoa!

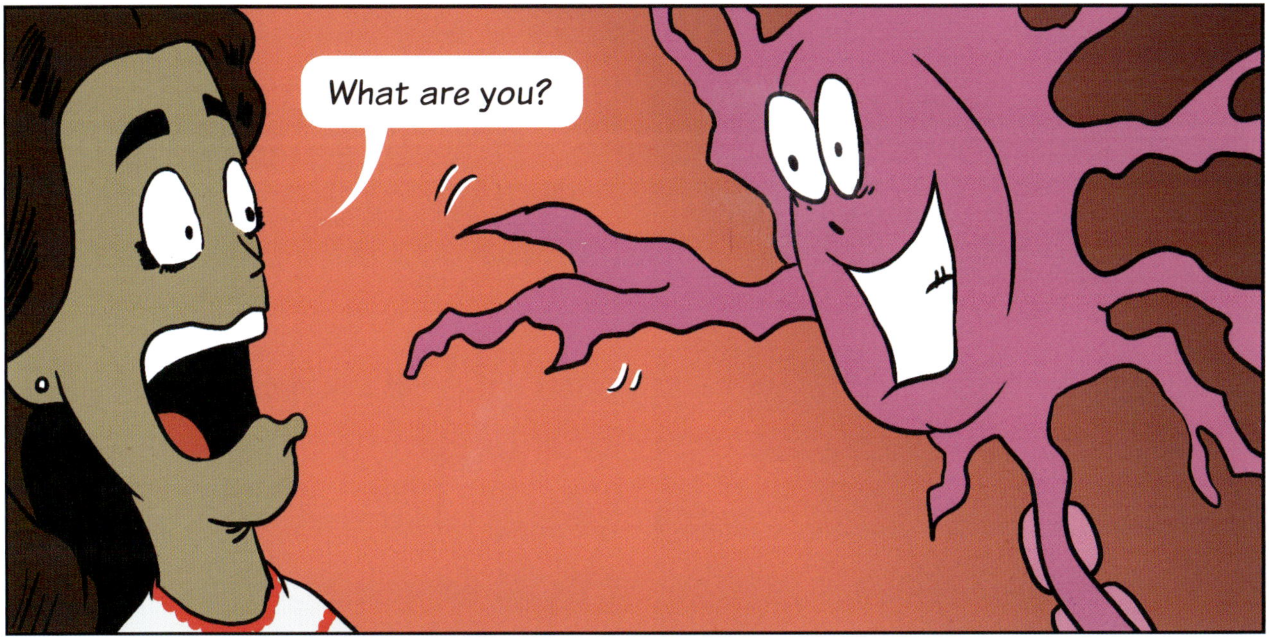
What are you?

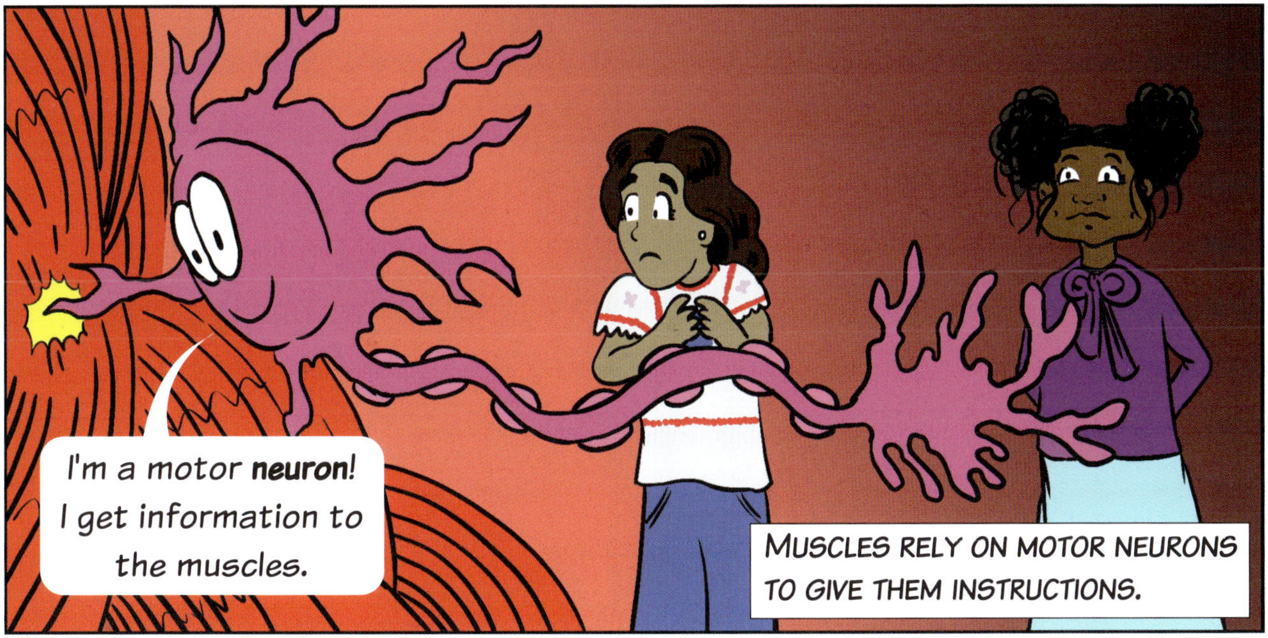
I'm a motor **neuron**! I get information to the muscles.
MUSCLES RELY ON MOTOR NEURONS TO GIVE THEM INSTRUCTIONS.

SMOOTH VS. CARDIAC

Both smooth muscles and **cardiac** muscles move involuntarily. That means a person does not have to think for them to move. Smooth muscle is found in organs such as the stomach and bladder. The **contractions** help those organs work for the body. Cardiac muscle is found only in the heart. It keeps the heart pumping.

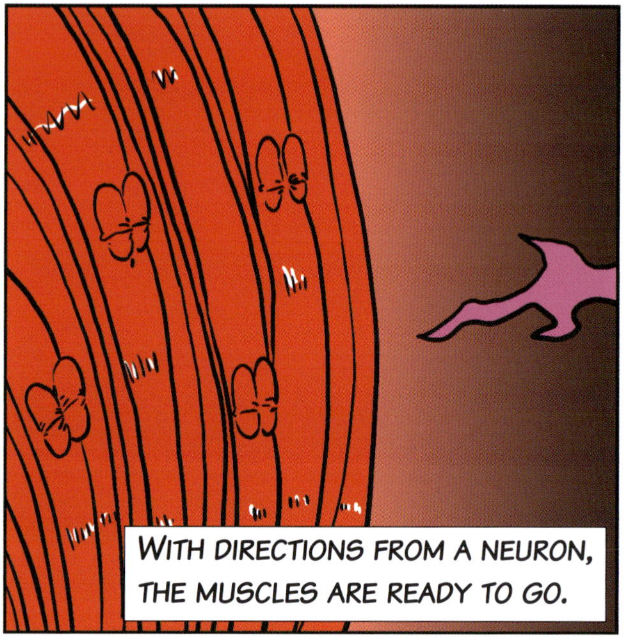

WITH DIRECTIONS FROM A NEURON, THE MUSCLES ARE READY TO GO.

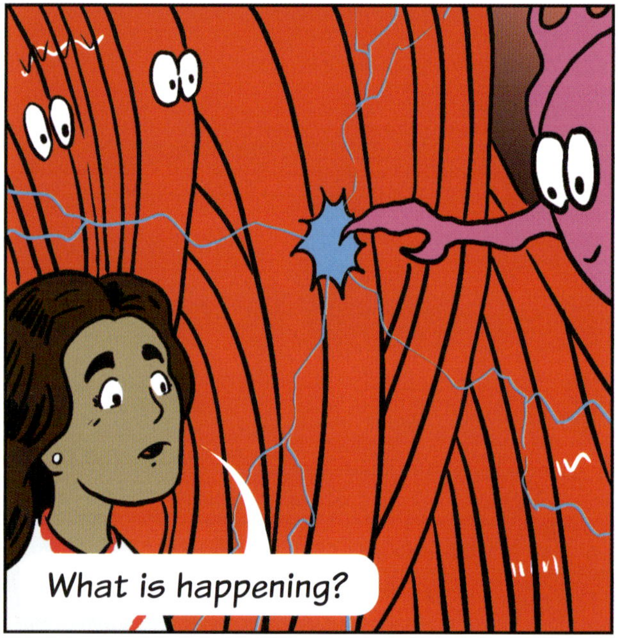

What is happening?

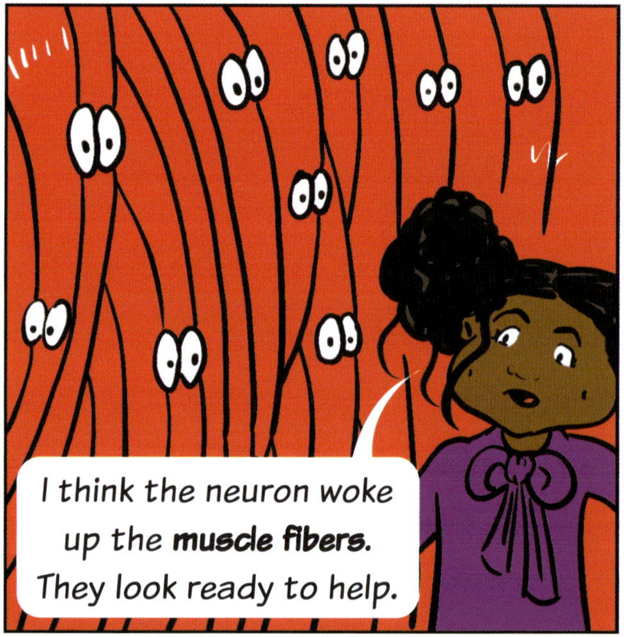

I think the neuron woke up the **muscle fibers**. They look ready to help.

Wow! Look at how many muscle fibers make up one muscle.

AS THE NEURON GETS THE SIGNAL TO THE MUSCLES, THE FIBERS ARE THEN READY FOR ACTION.

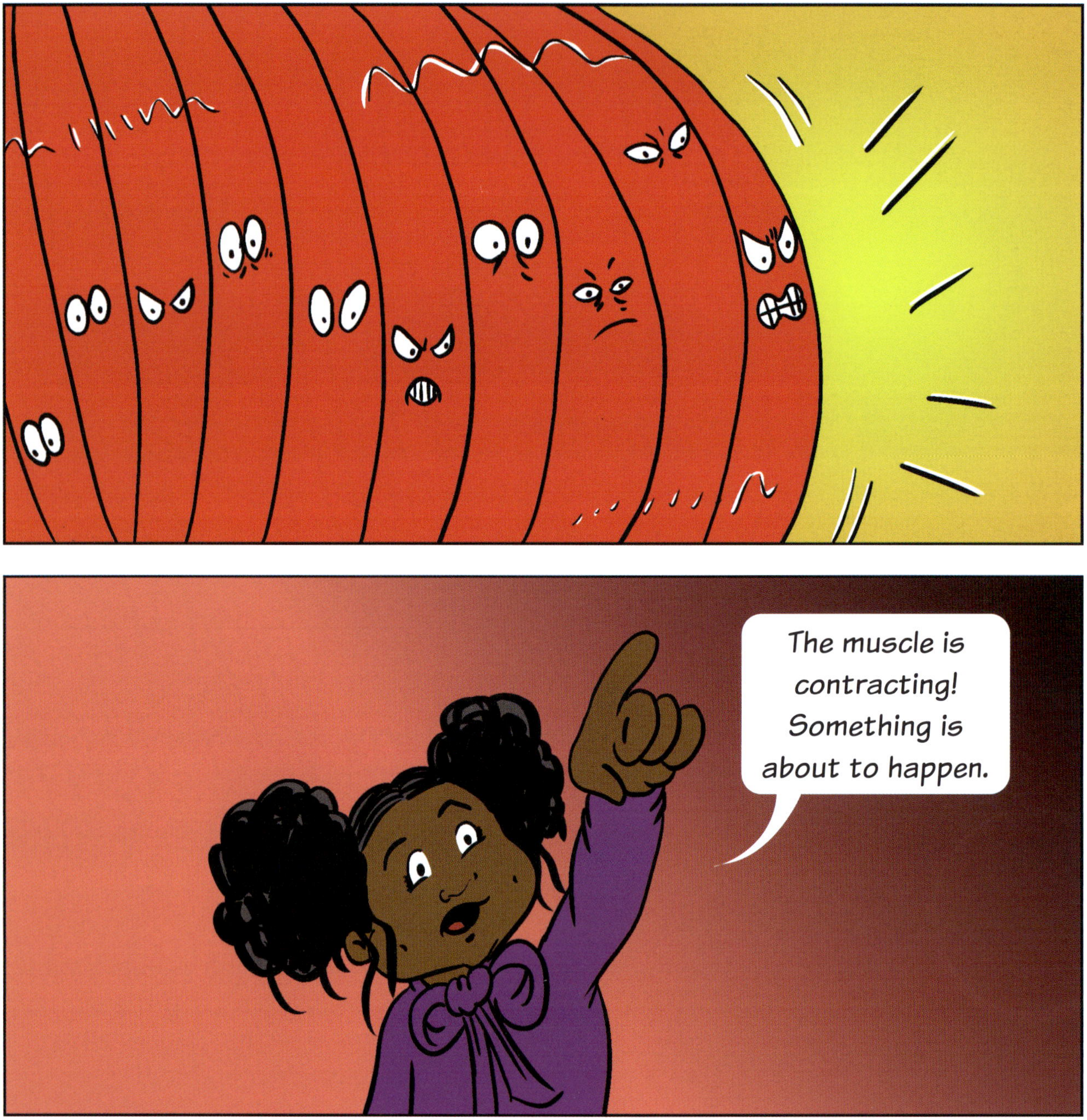
The muscle is contracting! Something is about to happen.

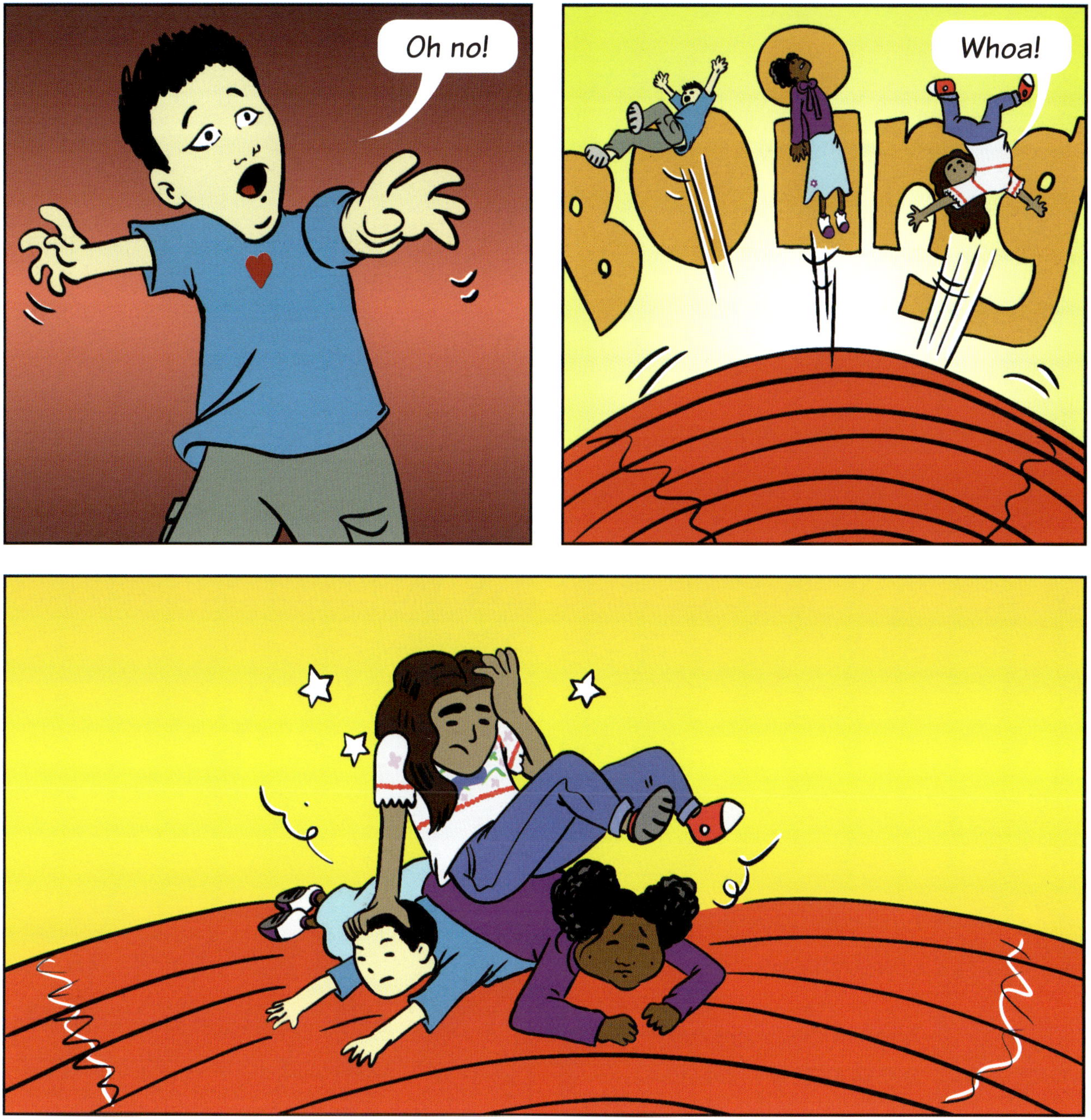
Oh no!
Boing
Whoa!

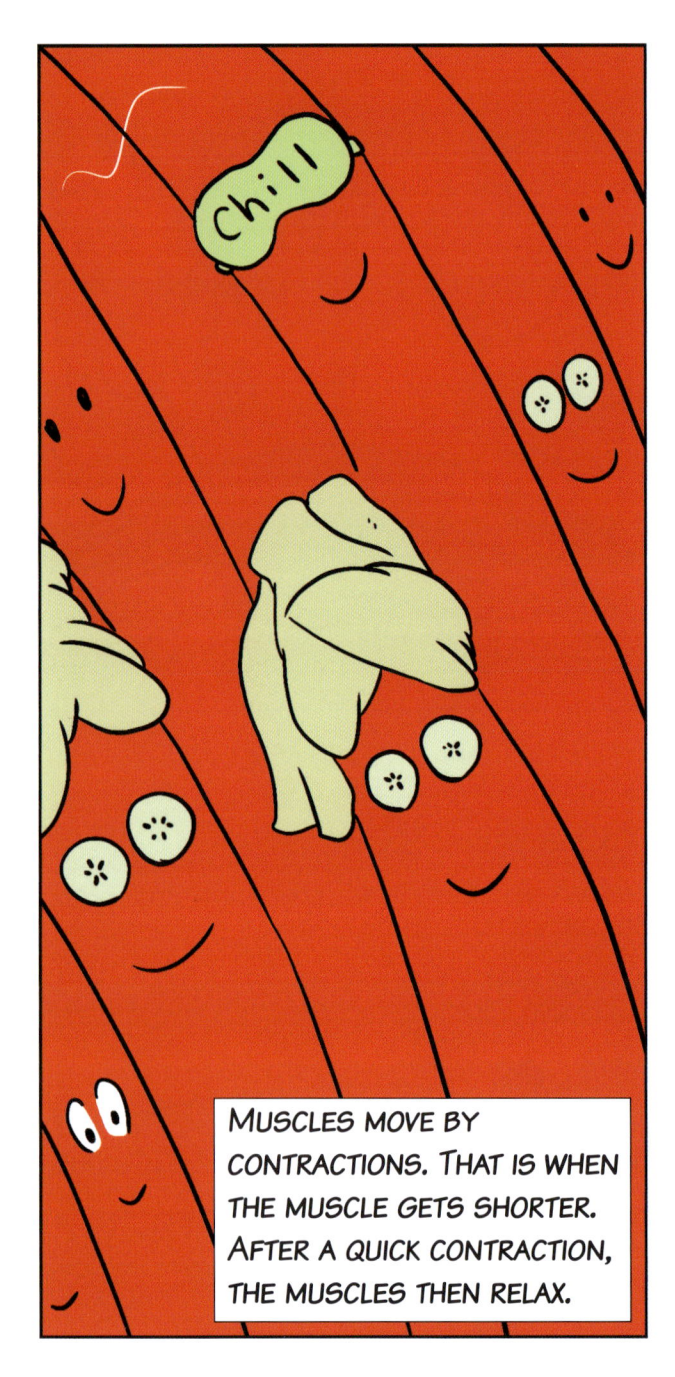

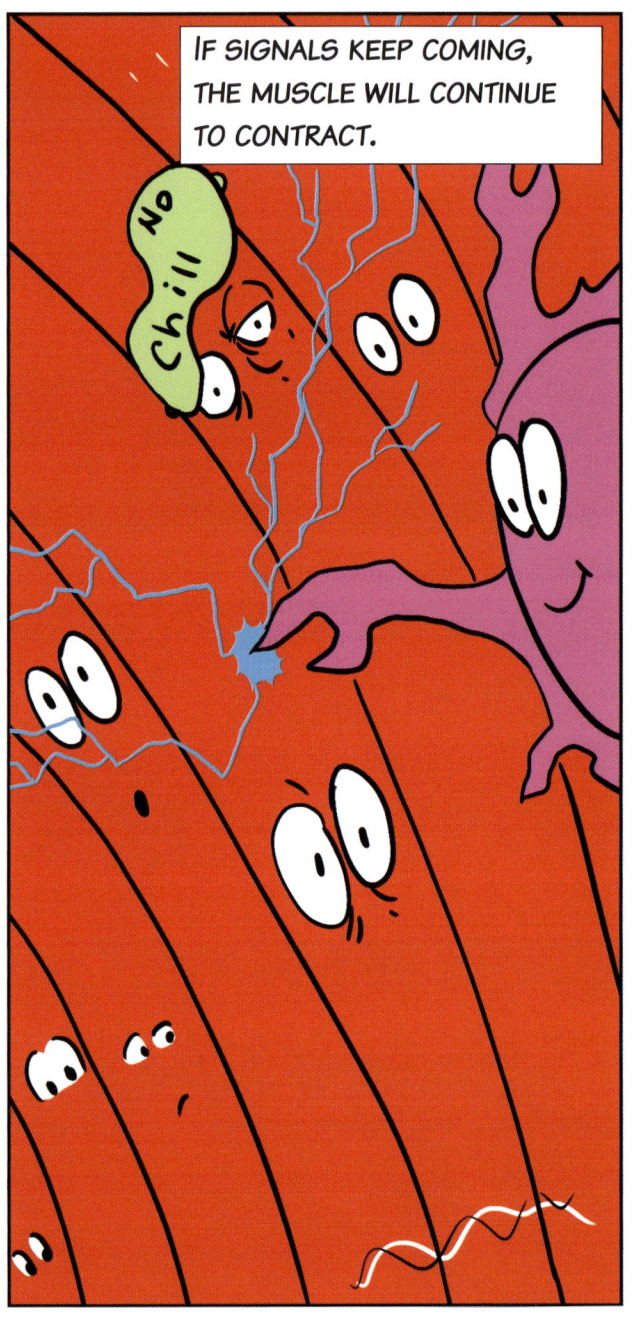

MUSCLES MOVE BY CONTRACTIONS. THAT IS WHEN THE MUSCLE GETS SHORTER. AFTER A QUICK CONTRACTION, THE MUSCLES THEN RELAX.

IF SIGNALS KEEP COMING, THE MUSCLE WILL CONTINUE TO CONTRACT.

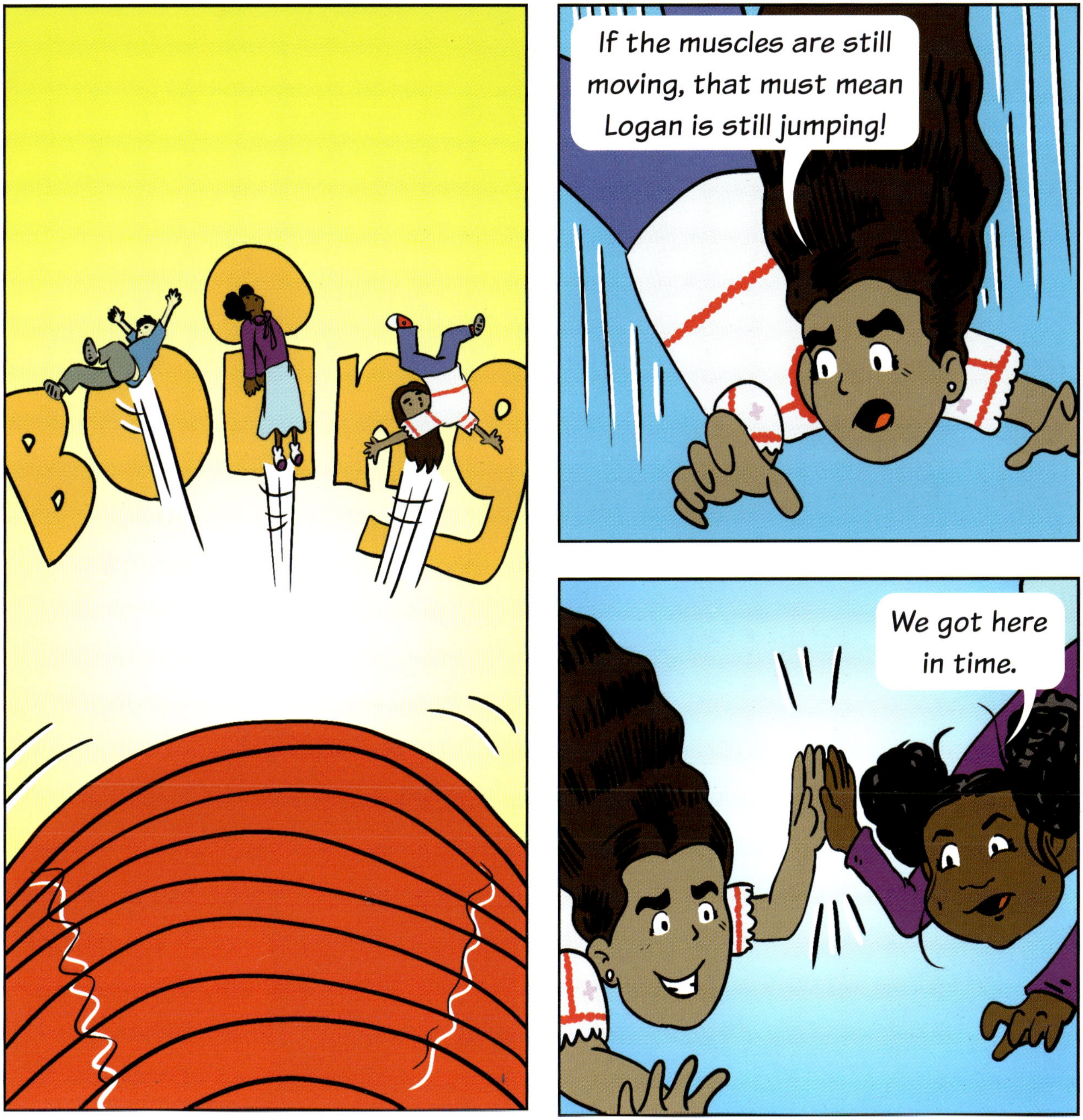
Boing
If the muscles are still moving, that must mean Logan is still jumping!
We got here in time.

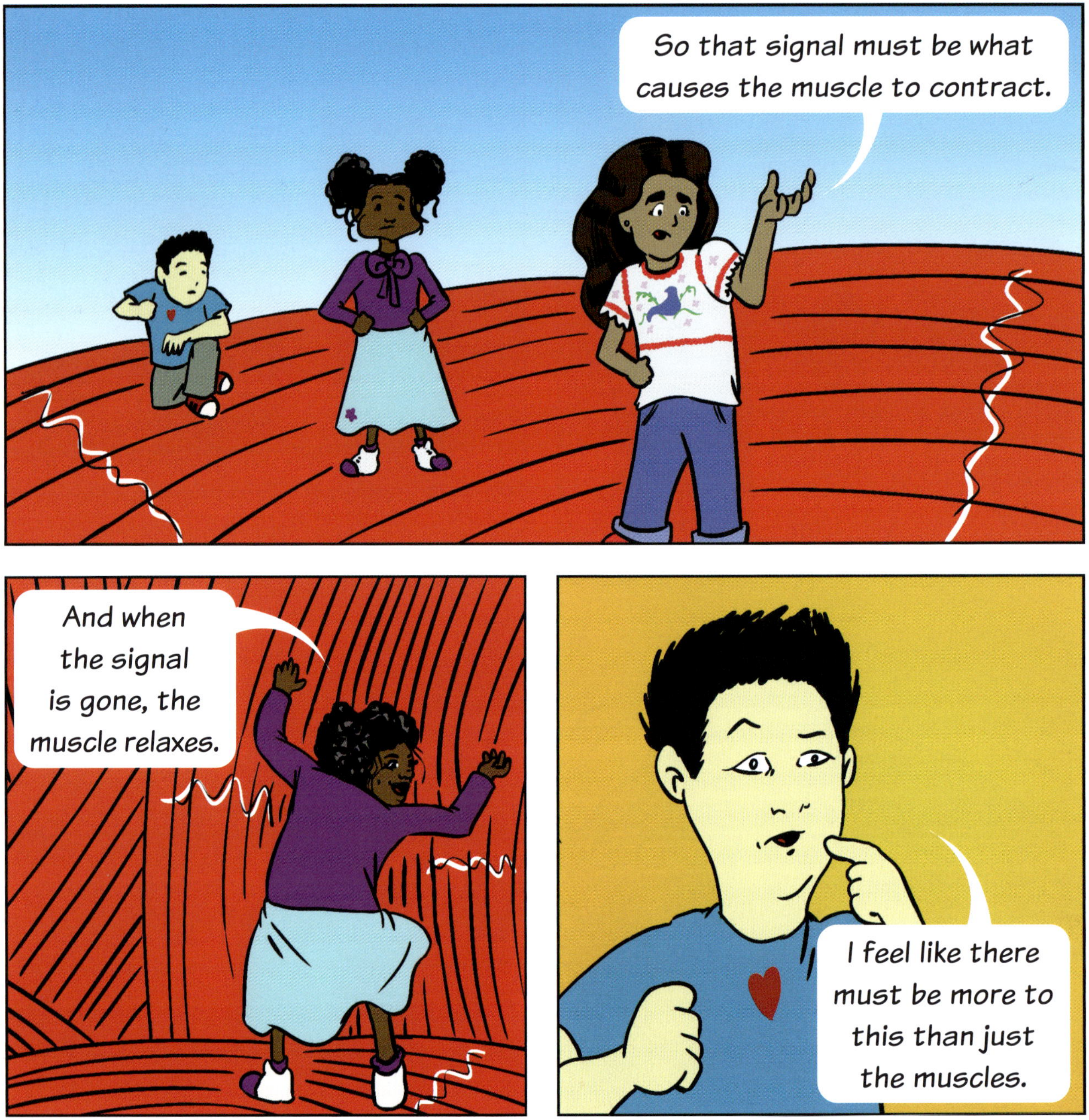
So that signal must be what causes the muscle to contract.
And when the signal is gone, the muscle relaxes.
I feel like there must be more to this than just the muscles.

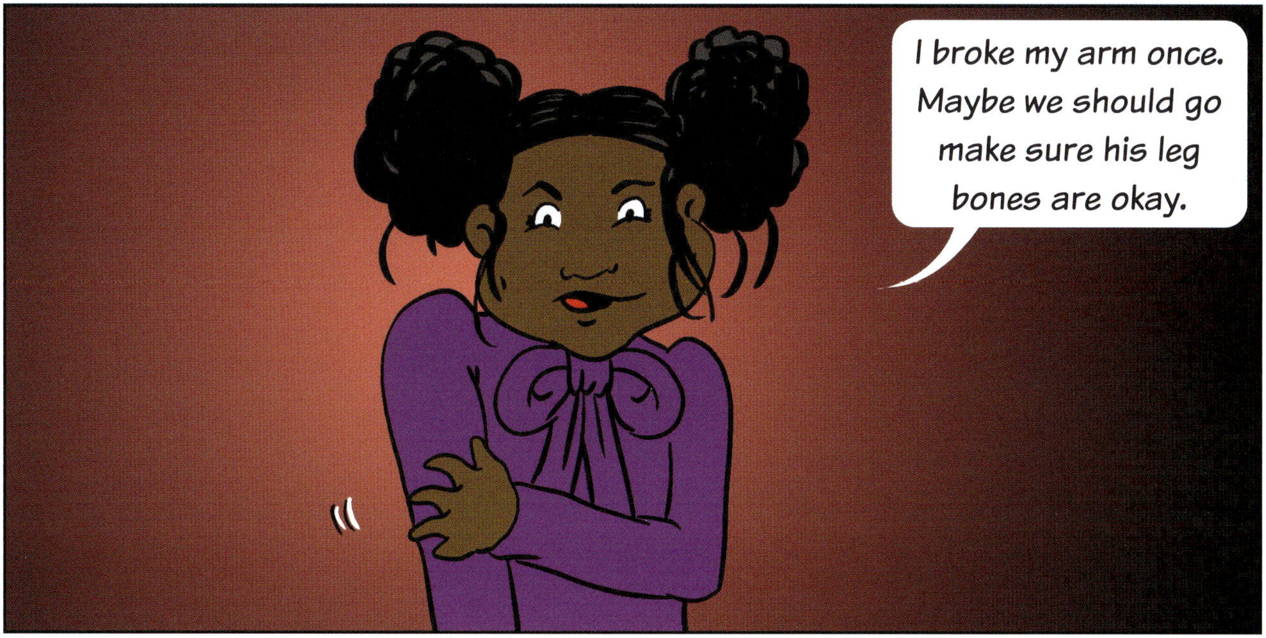
I broke my arm once. Maybe we should go make sure his leg bones are okay.

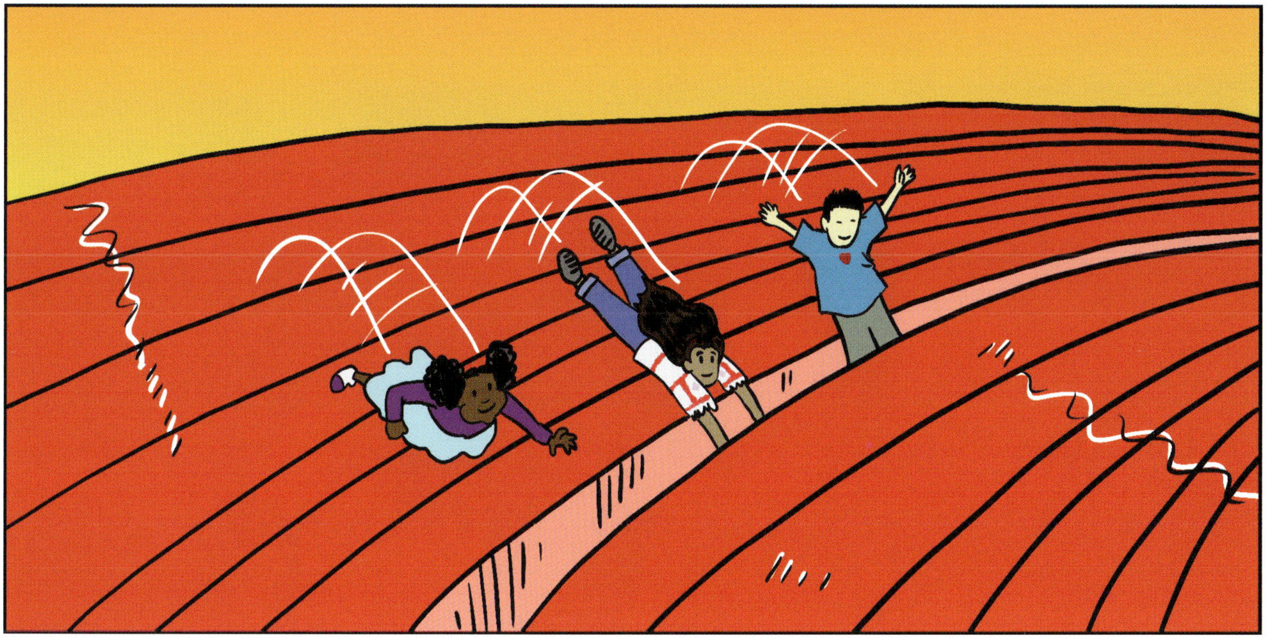

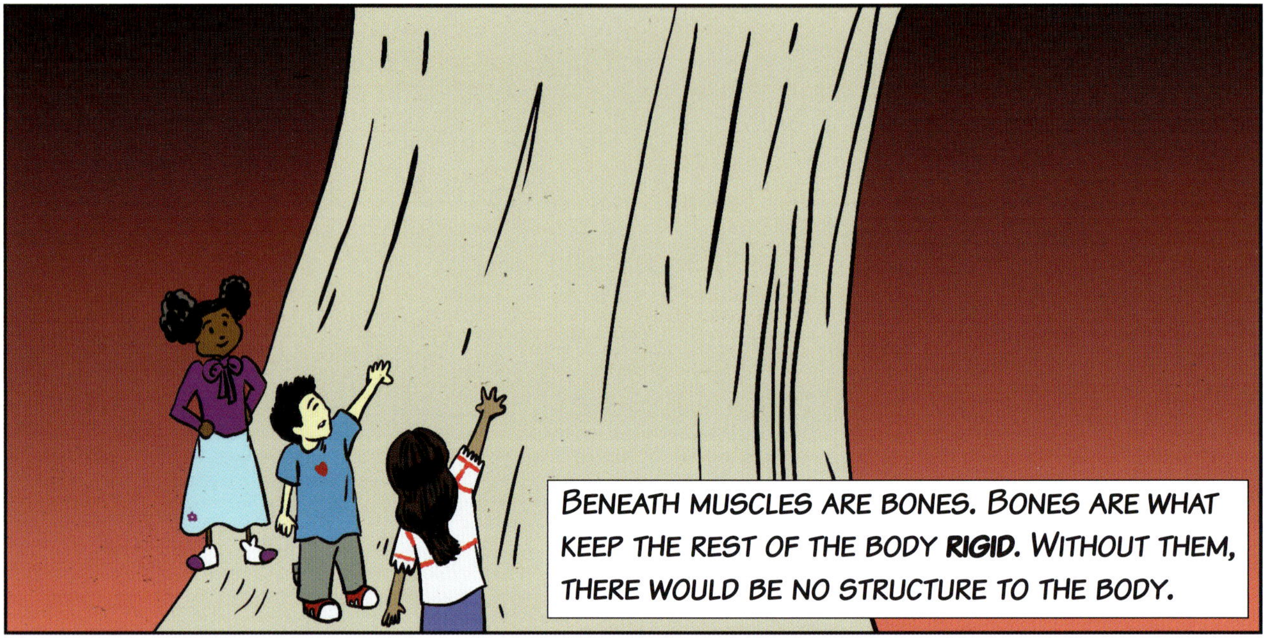
BENEATH MUSCLES ARE BONES. BONES ARE WHAT KEEP THE REST OF THE BODY **RIGID**. WITHOUT THEM, THERE WOULD BE NO STRUCTURE TO THE BODY.

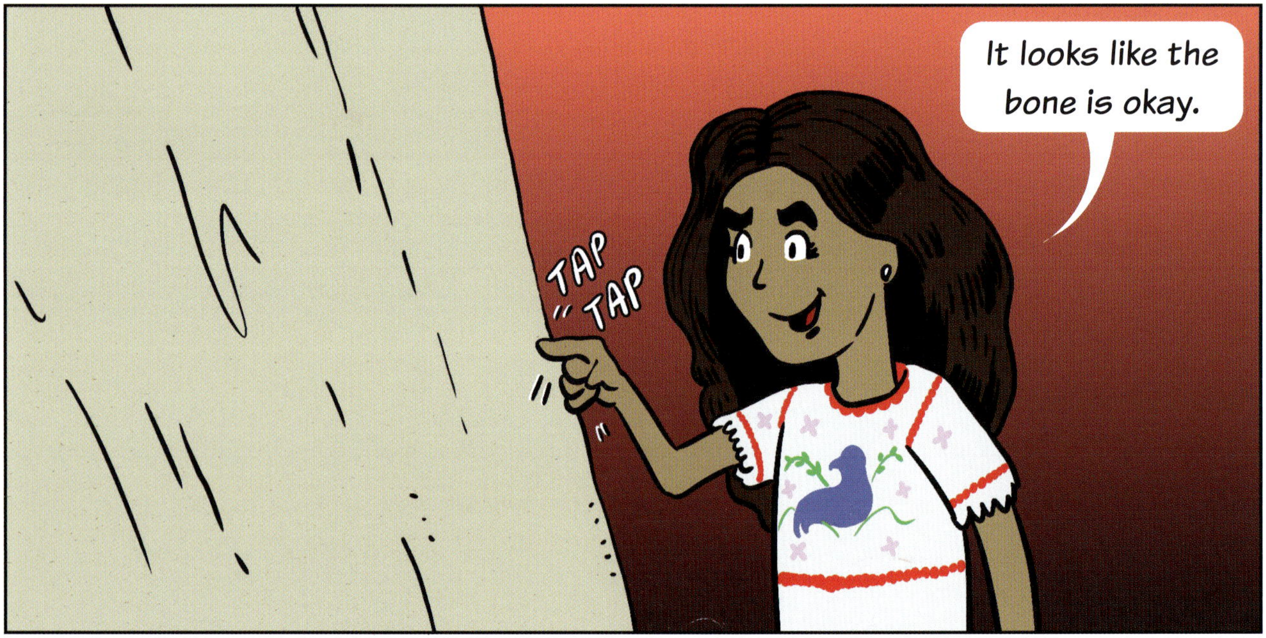
It looks like the bone is okay.
TAP TAP

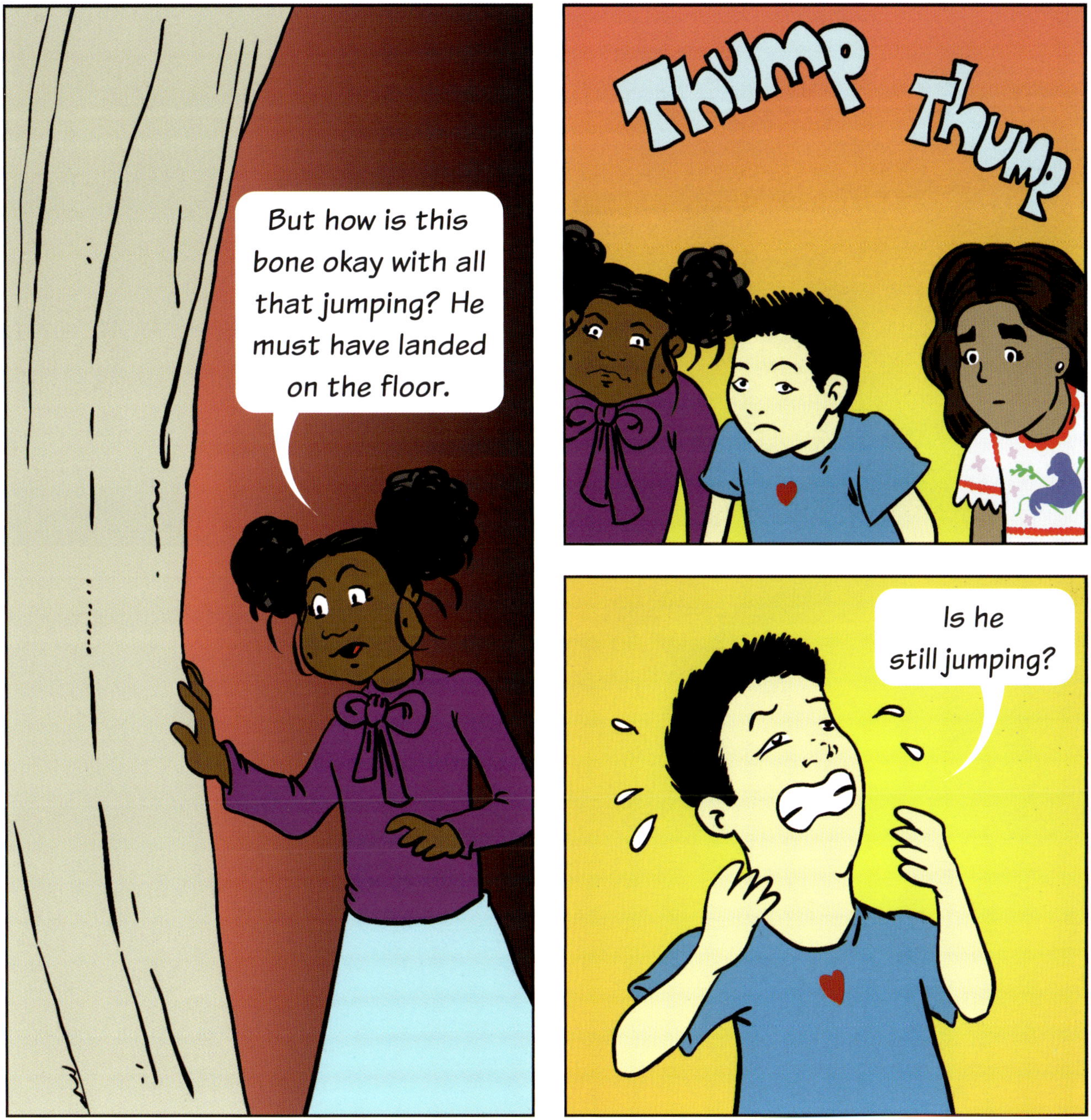
But how is this bone okay with all that jumping? He must have landed on the floor.
THUMP THUMP
Is he still jumping?

MUSCLE CONTRACTIONS

Acetylcholine (uh-set-ul-KOH-leen) is a very important chemical in the body. It is found in every motor neuron. It is what signals to muscles that it is time to contract. From a bicep curling to an eyelash blinking, all movement in the body is signaled by acetylcholine.

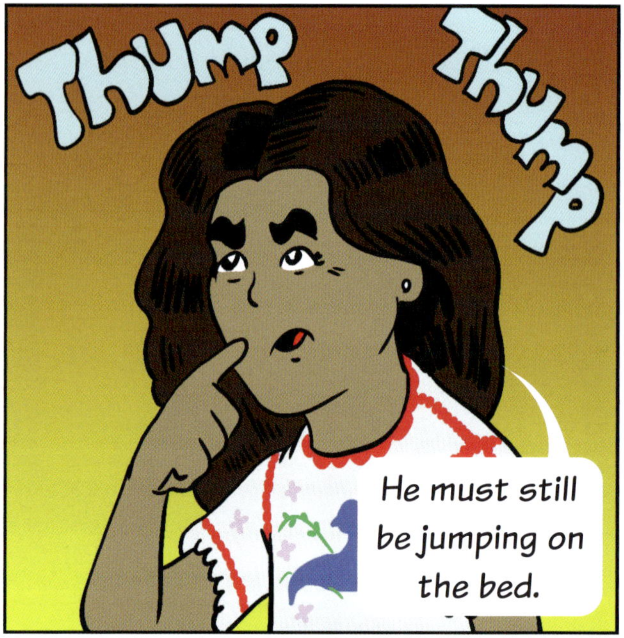
THUMP
THUMP
He must still be jumping on the bed.

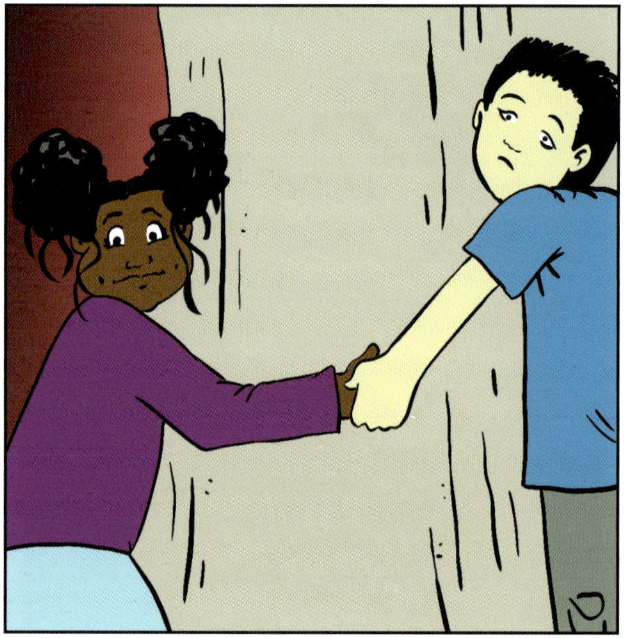

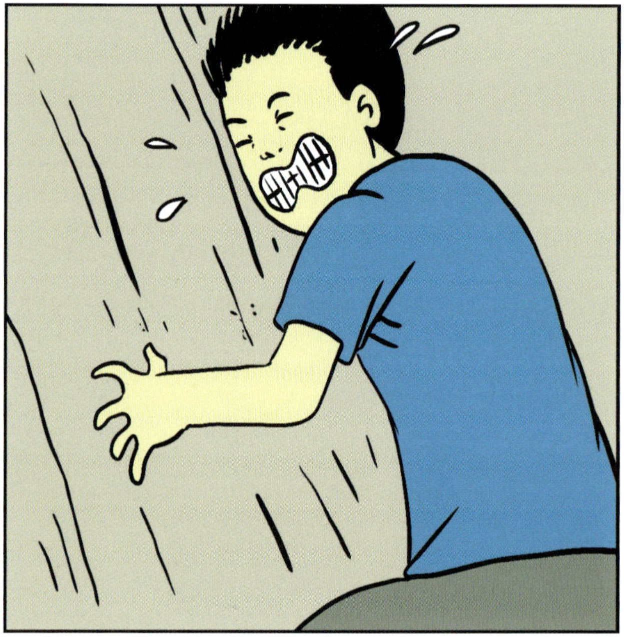

You don't need to hold on. Look, the muscles are taking the **impact**.
TAP

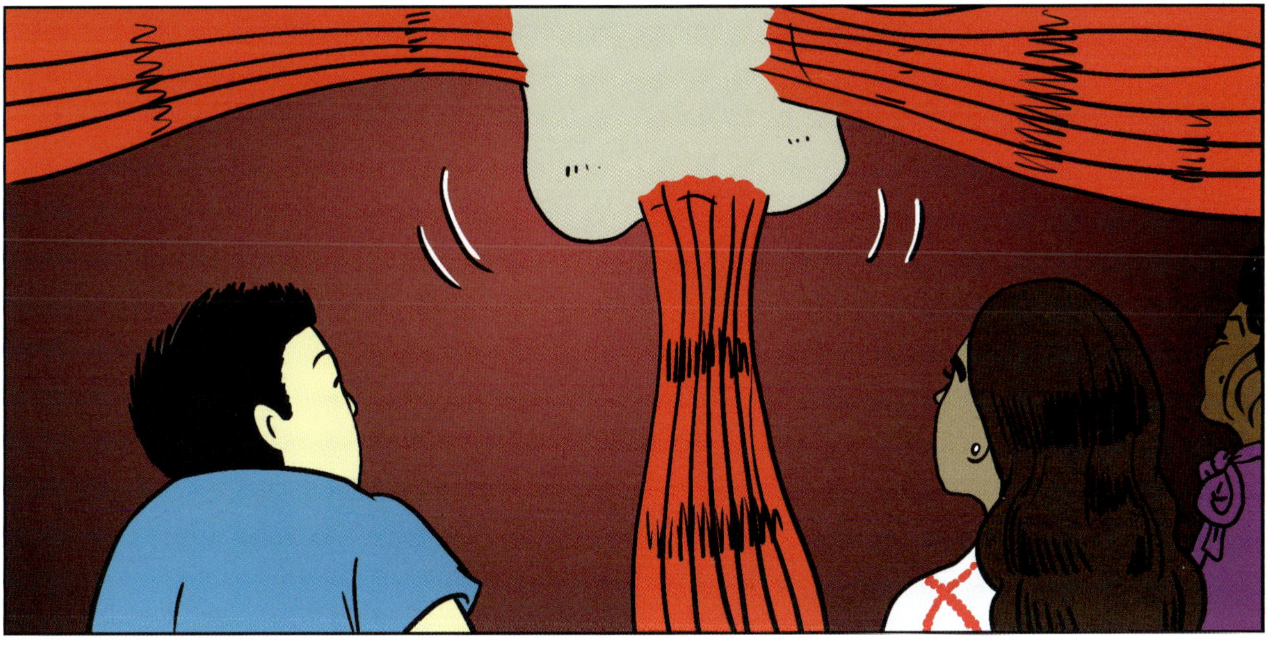

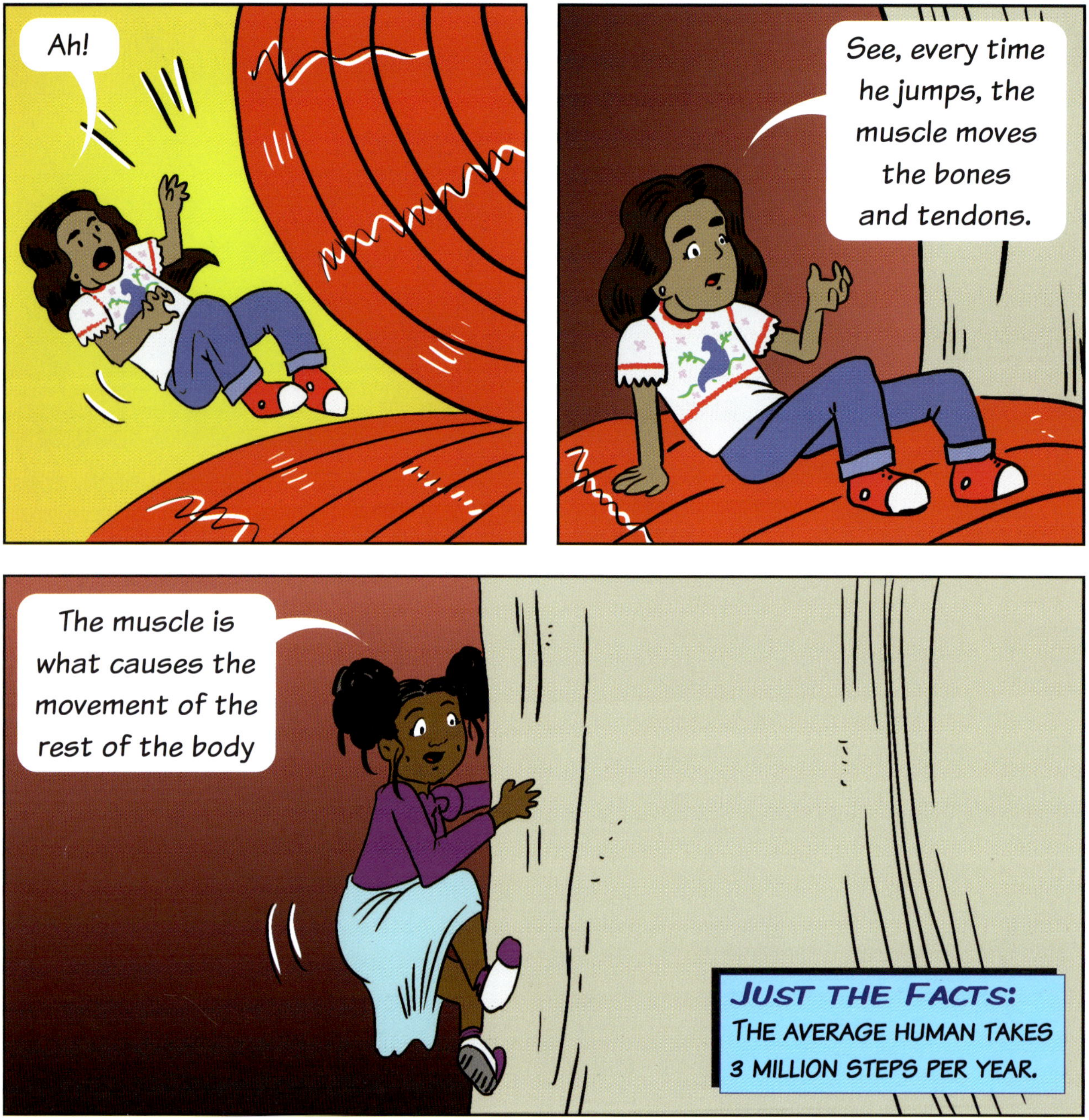
Ah!
See, every time he jumps, the muscle moves the bones and tendons.
The muscle is what causes the movement of the rest of the body
JUST THE FACTS:
THE AVERAGE HUMAN TAKES 3 MILLION STEPS PER YEAR.

And that must be a tendon. That's what connects the muscle to the bone.

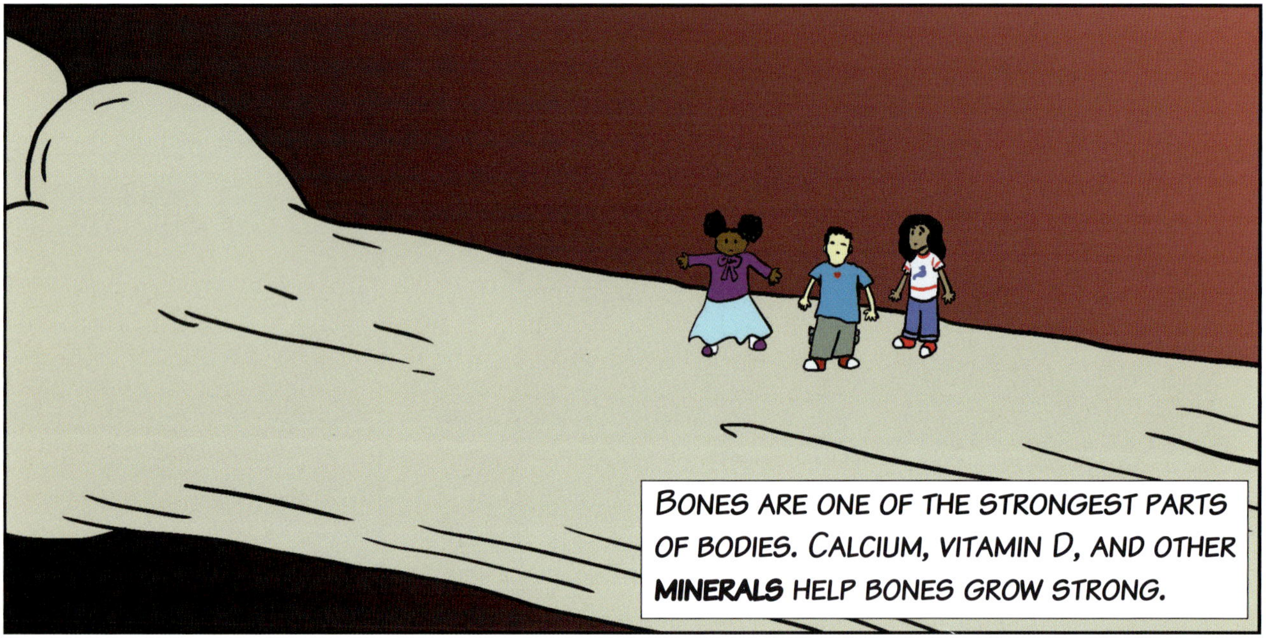
BONES ARE ONE OF THE STRONGEST PARTS OF BODIES. CALCIUM, VITAMIN D, AND OTHER **MINERALS** HELP BONES GROW STRONG.

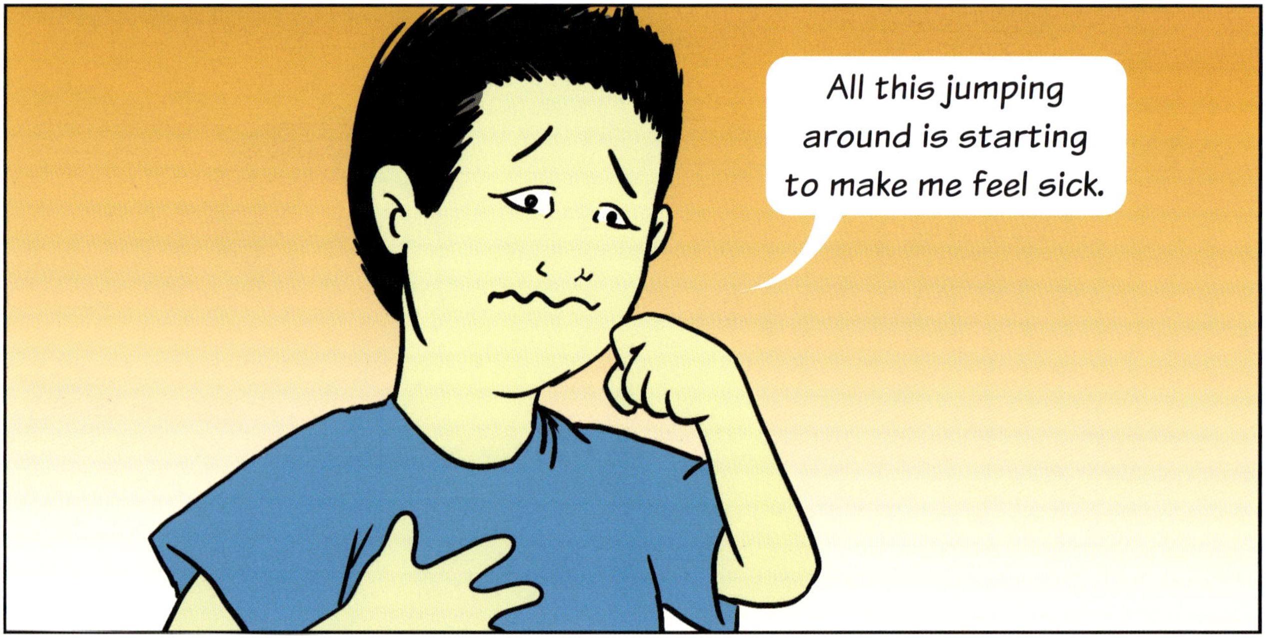
All this jumping around is starting to make me feel sick.

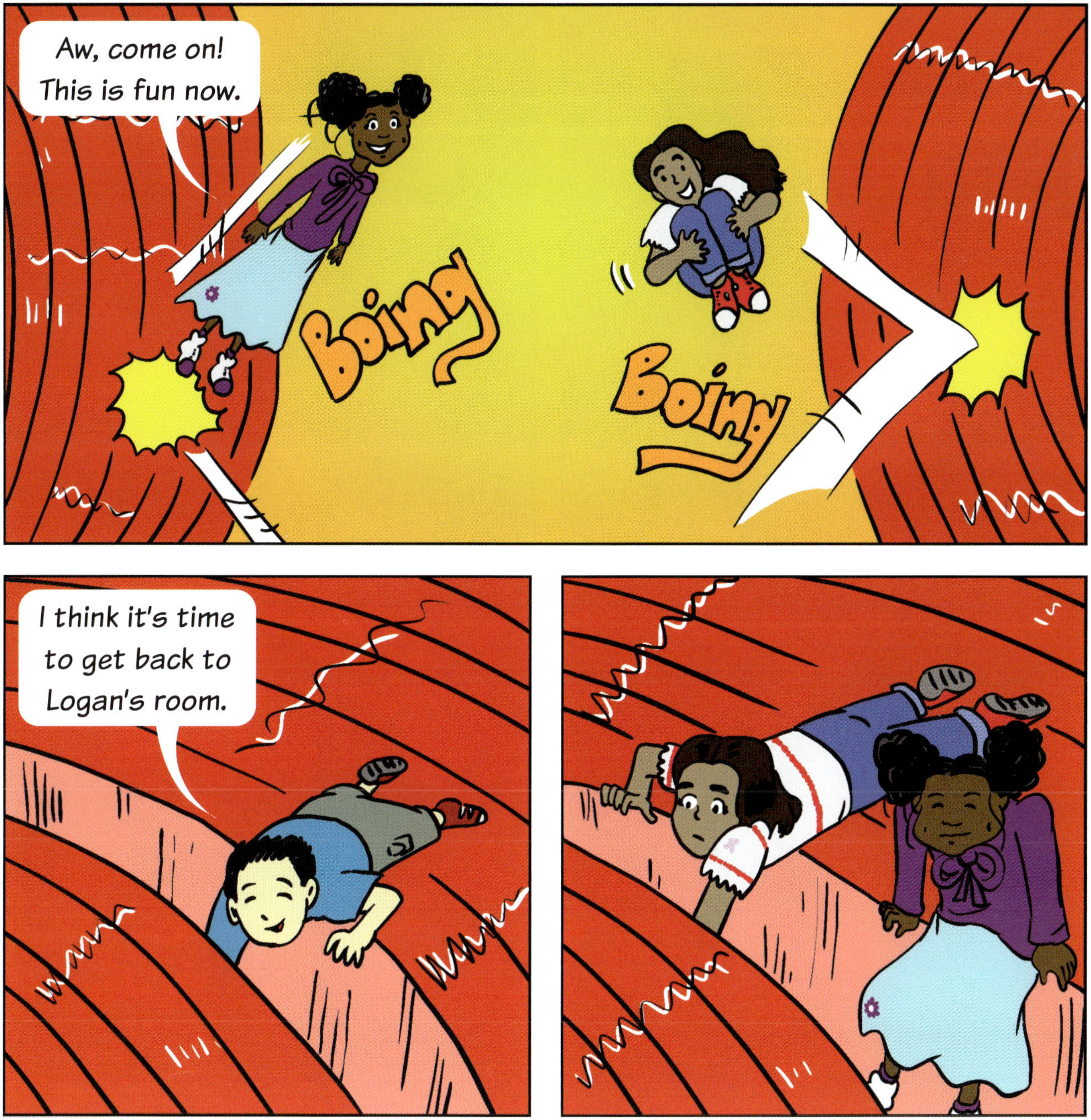
Aw, come on! This is fun now.
Boing
Boing
I think it's time to get back to Logan's room.

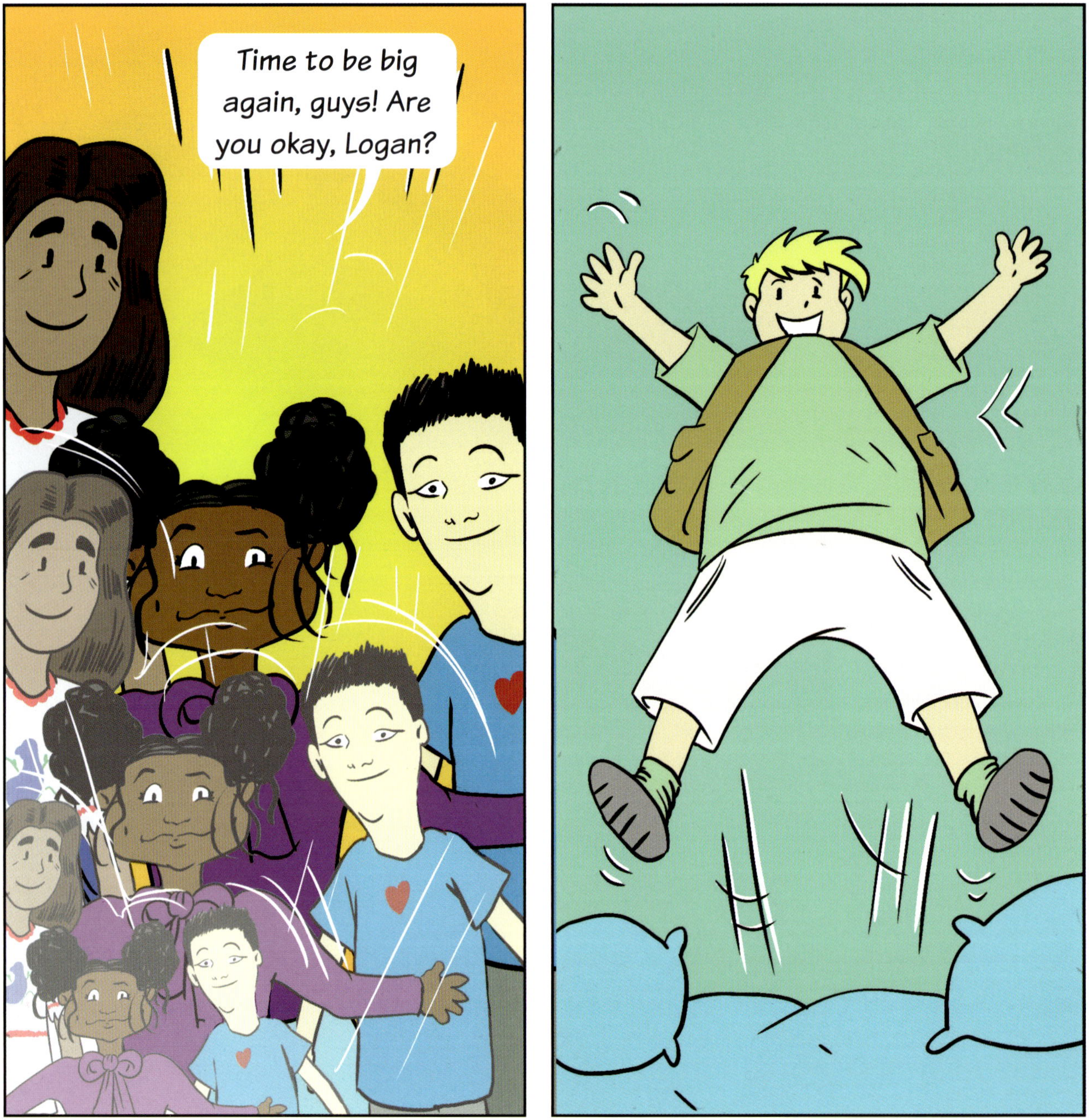
Time to be big again, guys! Are you okay, Logan?

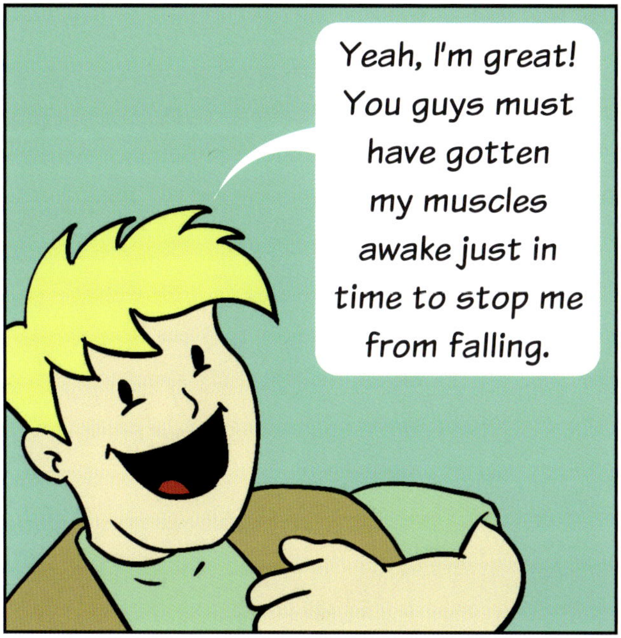
Yeah, I'm great! You guys must have gotten my muscles awake just in time to stop me from falling.

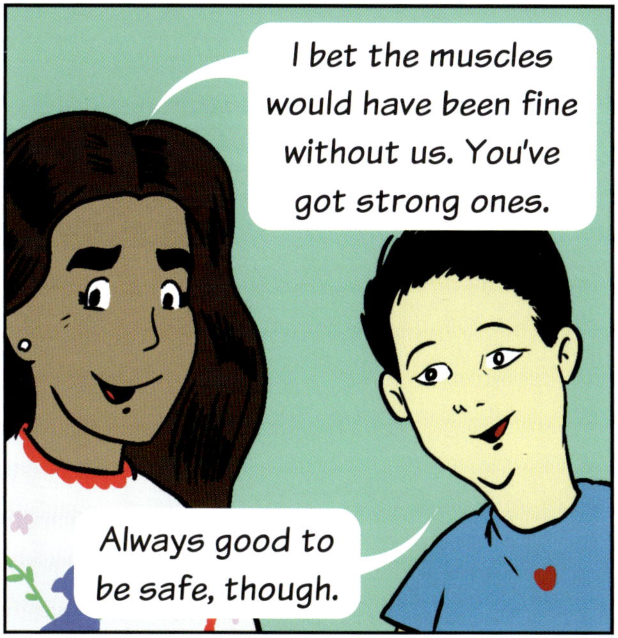
I bet the muscles would have been fine without us. You've got strong ones.
Always good to be safe, though.

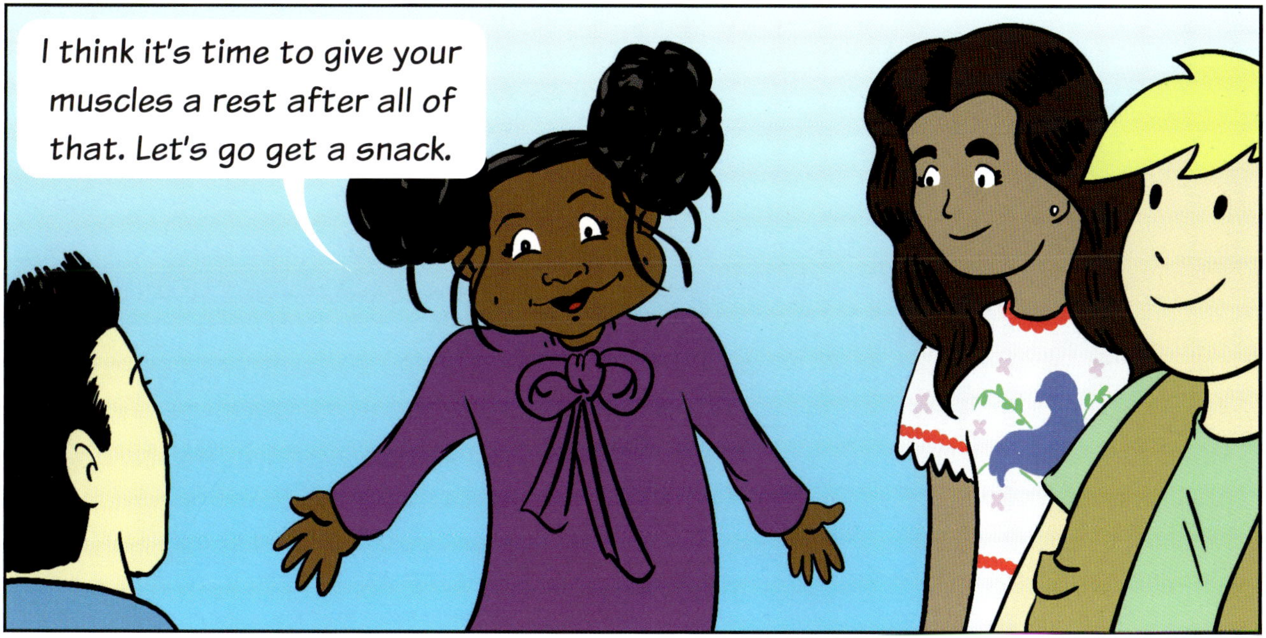
I think it's time to give your muscles a rest after all of that. Let's go get a snack.

The Muscular System

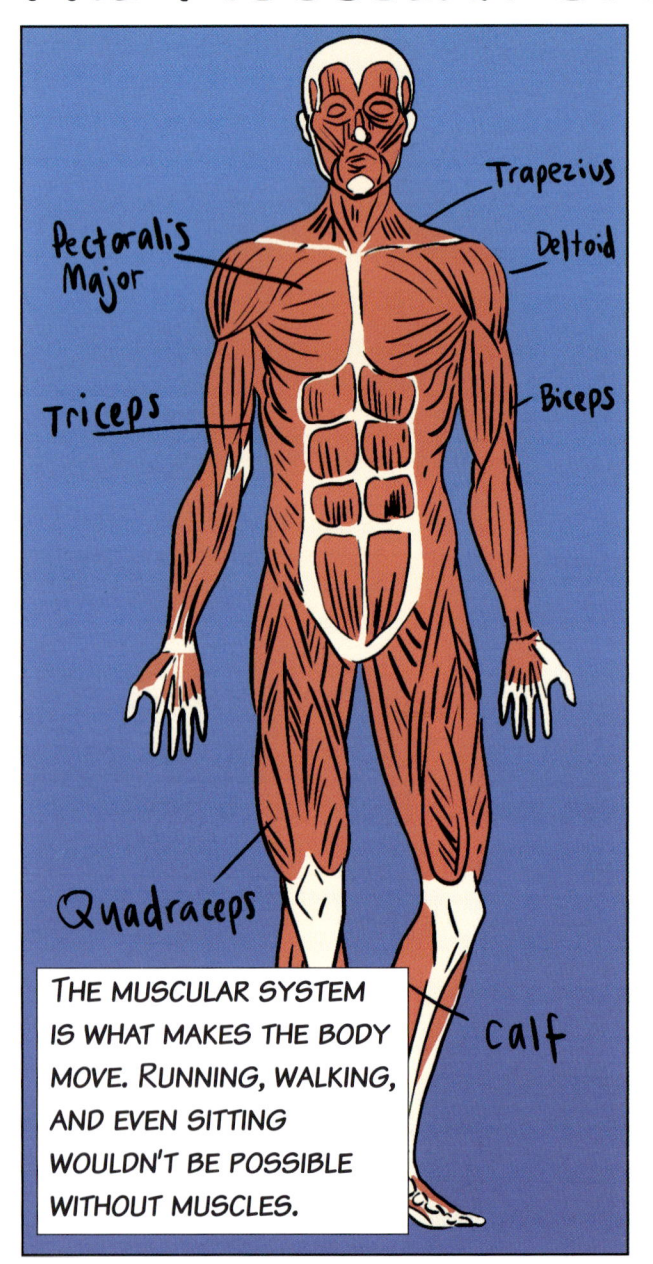

Trapezius
Deltoid
Pectoralis Major
Triceps
Biceps
Quadraceps
Calf

The muscular system is what makes the body move. Running, walking, and even sitting wouldn't be possible without muscles.

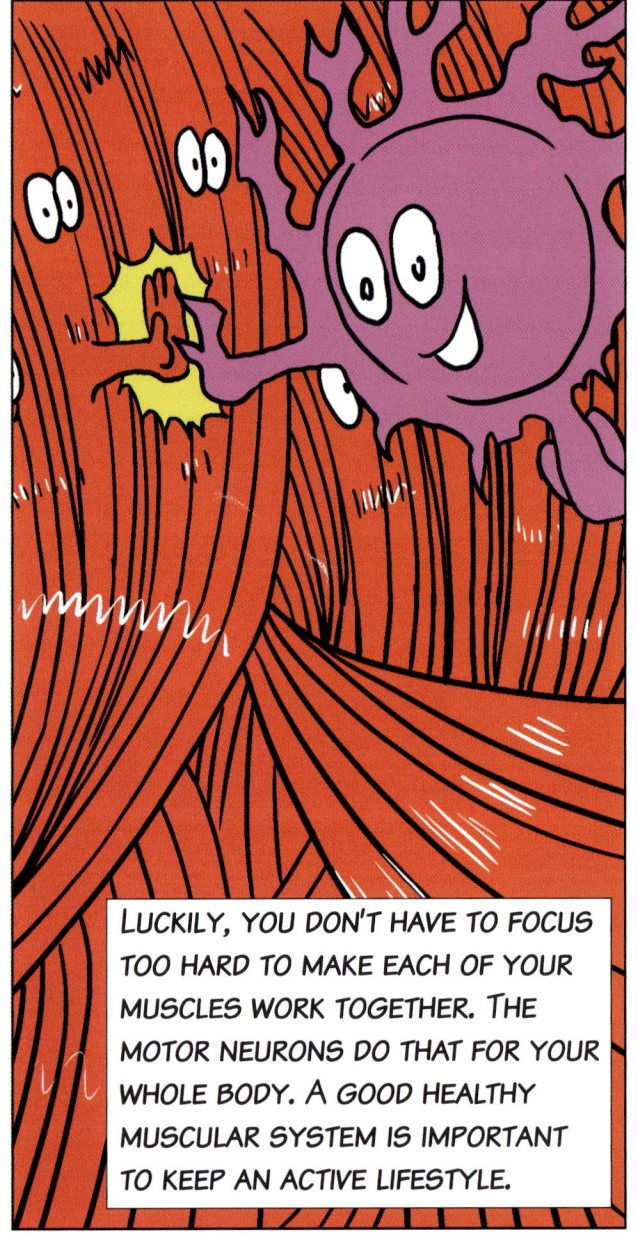

Luckily, you don't have to focus too hard to make each of your muscles work together. The motor neurons do that for your whole body. A good healthy muscular system is important to keep an active lifestyle.

Glossary

absorb: to take in or soak up in a slow and natural way

cardiac: having to do with the heart

contractions: movements of tightening up and becoming smaller

efficiency: the ability to do a task without much waste

impact: the force of two objects colliding

microscopic: too small to be seen without a tool, such as a microscope

minerals: substances that occur in nature and are important for bodily health

muscle fibers: bundles of ropey pieces of tissue that make up a muscle

neuron: a specialized cell that carries messages throughout the nervous system

rigid: stiff

Further Reading

Cole, Tayler. *20 Fun Facts about the Muscular System.* New York: Gareth Stevens Publishing, 2019. Learn fun facts about the most interesting parts of the human muscular system.

Huddleston, Emma. *Looking at Layers.* Mankato, MN: The Child's World, 2020. Explore the human body by looking at it in layers, from skin to muscles to skeleton.

Lawton, Cassie M. *The Human Muscular System.* New York: Cavendish Square Publishing, 2021. Read more about the major muscles in the body, what they do, and how to help them grow strong.

Anatomy Arcade: Muscular Jigsaw (http://www.anatomyarcade.com/games/jigsaws/MuscularJigsaw/muscularJigsaw.html) Play a puzzle game to learn the major muscles in the body.

KidsHealth: Your Muscles (https://kidshealth.org/en/kids/muscles.html) KidsHealth explains the importance of muscles and how to keep them healthy.

Visible Body: Muscular System Overview (https://www.visiblebody.com/learn/muscular/muscular-overview) View the muscular system in 3-D and see how it moves.

About the Author

Alexander Lowe is a writer who splits his time between Los Angeles and Chicago. He has written children's books about sports, technology, science, and media. He has also done extensive work as a sportswriter and film critic. He loves reading books of any and all kinds.

About the Illustrator

Sebastian Kadlecik is a screenwriter, actor, and comic book maker. He is best known as the creator of the epic action saga *Penguins vs. Possums*, about a secret, interspecies war for dominion over the earth, and the Eisner-nominated *Quince*, about a young Latina who gets superpowers at her quinceañera.